# Omics Approaches, Technologies And Applications

Preeti Arivaradarajan • Gauri Misra

Editors

# Omics Approaches, Technologies And Applications

Integrative Approaches For Understanding OMICS Data

 Springer

*Editors*
Preeti Arivaradarajan
Amity Institute of Biotechnology
Amity University
Noida, Uttar Pradesh, India

Gauri Misra
Amity Institute of Biotechnology
Amity University
Noida, Uttar Pradesh, India

ISBN 978-981-13-2924-1     ISBN 978-981-13-2925-8   (eBook)
https://doi.org/10.1007/978-981-13-2925-8

Library of Congress Control Number: 2018965925

This Springer imprint is published by the registered company Springer Nature Singapore Pte Ltd.
The registered company address is: 152 Beach Road, #21-01/04 Gateway East, Singapore 189721, Singapore

*Dedicated to*
*The lotus feet of Shri Radha Krishna*
*and*
*Goverdhannathji*

# Foreword

This book examines omics technologies and introduces the subject for new readers. It is particularly suitable for students (graduate, postgraduate, and doctoral), analytical scientists, and lab technicians. However, it has enough information to allow established scientists in related fields to understand the power and limitations of omics technology.

Omics is a relatively new area of study and cuts across all the biological disciplines and is relevant to all biological sciences. It attempts to look at biological systems in a holistic way and to account for all the interactions between, genes, proteins, RNA, and metabolites. It is divided into various disciplines which are discussed in the book. The chapters span from genomics that studies the structure and function of the whole genomes of organisms, proteomics describing the expressed proteins in a cell or tissue, transcriptomics dealing with the RNA present in a cell or tissue leading to an understanding about differential gene expression under particular conditions to metabolomics which provides a glimpse of the end products of metabolism contributing most to phenotypes. The book also contributes toward an in-depth understanding of the microbiome that encompasses the total genes of all the microorganisms in a particular ecological niche. Humans have their own microbiome (as do almost all environments on Earth), and there are more microbial cells in a human than there are human cells. The microbiome has gathered a lot of scientific attention recently for exploitation of its therapeutic relevance in humans, other animals, and plants; thus, it is sometimes considered as a separate discipline.

The book is edited by two exciting young scientists, one of whom I had the privilege of supervising for part of her PhD. They have invited the experts in the thematic areas to design a contributed volume delving into different omics branches.

Eastman Dental Institute
University College London
London, UK

Peter Mullany

# Preface

Our exposure to the omics world during doctoral program in India and abroad laid the foundation for this book. The advent of highly parallel assays led to the transition of biological research from discrete knowledge of gene/transcript/protein/metabolite to a complete interlinked biological picture. This field has gained momentum with its immense usefulness across various dimensions including disease prognosis, therapeutics, personalized medicine, and drug discovery. Therefore, it is imperative to design a resource that will not only be useful for beginners but also for the experts seeking the advancement of their knowledge in this field.

The present book is divided into seven chapters with an aim to address the fundamental questions of diverse audience pertaining to the interdisciplinary field of omics. The introductory chapter describes the scope of omics, experimental design in omics research, its applications, and the usage of R language for analyzing high-throughput omics data. The second chapter outlines a coherent view of the human genome architecture, DNA sequencing approaches, and new technological advances in genomics. The third chapter discusses the principle of transcriptomics, technologies (expression sequence tag, serial/cap analysis of gene expression, microarray, RNA-seq) used to study transcriptomes, and applications of transcriptomics in disease profiling, ecology, evolution, and gene function annotation. The fourth chapter details different types of proteomics and advanced proteomic techniques such as two-dimensional electrophoresis, isotype-coded affinity tag peptide labeling, mass spectrometry, and multidimensional protein identification technique. The fifth chapter describes metabolome, its applications, and integrated platforms for analysis and interpretation of the metabolomics data. The sixth chapter provides an insight to soil, plant, marine, and human microbiome. A special emphasis is laid on human coinhabitants, wherein microbiome of various niches such as the gut, skin, oral, and urine is discussed in detail. The last chapter on bioinformatics resources gives an in-depth description about various bioinformatics approaches available to analyze genomics, transcriptomics, proteomics, and metabolomics data.

We believe our effort will be a priceless treasure for the general audience. The text has been enriched with the help of appropriate annotations, tables, and further readings. A positive feedback and scientific appreciation will be the true reward that the editors genuinely seek.

Noida, Uttar Pradesh, India                                            Preeti Arivaradarajan
                                                                                  Gauri Misra

# Contents

# Editors and Contributors

## About the Editors

**Preeti Arivaradarajan** is Assistant Professor at Amity University, Noida, Uttar Pradesh, India. Previously, she completed Ph.D. from the School of Biological Sciences, Madurai Kamaraj University, Tamil Nadu, India. Her research inclination is towards the study of microbiome associated with pristine environments, with a special interest for human oral microbiome. She has been conferred with various prestigious awards, notably, Commonwealth Split-Site Doctoral Scholarship by Commonwealth Scholarship Commission in United Kingdom, Innovation in Scientific Pursuit for Inspired Research (INSPIRE) fellowship by Department of Science and Technology, Government of India, and gold medal for the best outgoing student of the year during master's program by Sri Ramachandra University, Chennai, India. Furthermore, she is a member of the Indian Science Congress Association, and is serving as a reviewer for numerous international journals. She has also published a number of research articles in peer-reviewed international journals and has authored book chapters.

**Gauri Misra** is currently working as Assistant Professor at Amity University, Noida, Uttar Pradesh (UP), India. Before joining Amity University, she worked as Assistant Professor at Hygia Institute of Pharmaceutical Education and Research, Lucknow (UP), India. From 2010 to 2011, she worked as a postdoctoral fellow at CHUL Research Centre, Quebec, Canada, where she worked towards understanding the role of androgen receptor in the growth and proliferation of breast cancer cells using various structural biology approaches. During her doctoral studies at the Central Drug Research Institute, Lucknow, she made an innovative contribution towards understanding the structural and functional characterization of the *Plasmodium falciparum* proteins that are involved in the transit peptide-mediated pathway. Her research interests are in the field of pathogen biology and cancer. She is striving for a healthy future for mankind through her research efforts in the field of structure based drug design paving way for drug discovery. She is the recipient of many awards, including the Young Scientist best scientific presentation award at the International Conference on "Trends in Biochemical and Biomedical Research," Banaras Hindu University, Banaras, India, in 2018, the prestigious Eli-Lilly Asia Outstanding Thesis Award (first prize) in 2009, and best oral presentation award at the "National Seminar on Crystallography-37," held in Jadavpur University, Kolkata, in 2008. She has been an outstanding performer, receiving gold medals and honors at various stages of her academic journey. Previously, she was selected as visiting scientist under the INSA bilateral exchange program to visit the Israel Structural Proteomics Center situated at Weizmann Institute of Science, Rehovot, Israel, in 2014.

She is serving as the reviewer for various renowned international journals. Furthermore, she is a member of many scientific societies, including the Indian Biophysical Society, the Indian Science Congress Association, and the Indian Crystallographic Association. To date, she has authored and co-authored 13 articles in various peer-reviewed journals. She has successfully edited a book on biophysics with Springer in 2017. During the past 8 years, she has been actively involved in both research and teaching graduate and postgraduate students.

## Contributors

**Sudeep Bose**  Amity University, Noida, Uttar Pradesh, India

**Sangeeta Choudhury**  Research Department, Sir Ganga Ram Hospital, New Delhi, India

**Manali Datta**  Amity Institute of Biotechnology, Amity University Rajasthan, Jaipur, Rajasthan, India

**Neetu Jabalia**  Amity Institute of Biotechnology, Amity University, Noida, Uttar Pradesh, India

**S. V. Kirthanashri**  Amity Institute of Molecular Medicine & Stem Cell Research, Amity University, Noida, Uttar Pradesh, India

**Priyanka Narad**  Amity Institute of Biotechnology, Amity University, Noida, Uttar Pradesh, India

**Debarati Paul**  Amity University, Noida, Uttar Pradesh, India

**Jyotika Rajawat**  Molecular and Human genetics laboratory, Department of Zoology, University of Lucknow, Lucknow, Uttar Pradesh, India

**Abhishek Sengupta**  Amity University, Noida, Uttar Pradesh, India

**Desh Deepak Singh**  Amity Institute of Biotechnology, Amity University Rajasthan, Jaipur, Rajasthan, India

**Vivek Tanavde** Bioinformatics Institute, Agency for Science Technology and Research (A*STAR), Singapore, Singapore

Division of Biological and Life Sciences, School of Arts and Sciences, Ahmedabad University, Ahmedabad, India

**Candida Vaz** Bioinformatics Institute, Agency for Science Technology and Research (A*STAR), Singapore, Singapore

# Chapter 1
# Introduction to Omics

**Priyanka Narad and S. V. Kirthanashri**

**Abstract** Omics technologies also referred as high-dimensional biology encompasses the cells, tissues, and organisms in a manner that integrates the data from various platforms and helps in its interpretation. It primarily detects the genes (genomics), mRNAs (transcriptomics), proteins (proteomics), and metabolites (metabolomics) in a nontargeted and non-biased manner. The integration and interrelationships between networks of biological processes is termed as systems biology. The approach provides hope for unravelling the intricate details in various aspects of biology and accelerates innovation in healthcare. Understanding the various dimensions encompassing not only the three levels constituting the central dogma of life but also the intermediate metabolites is significant for the scientists to cover new horizons in drug discovery and disease regulation. This chapter outlines the scope of omics, experimental design in omics research, and its applications. It will also provide an overview to the usage of languages like R for analyzing high-throughput data from all branches of "omics" technologies.

The primary focus is to understand omics approaches that enable the validation of large-scale data that is generated from various experimental platforms. Systems biology and omics data are way apart from hypothesis-driven traditional studies. The systems biology experiments generate hypothesis by employing all data that needs to be further analyzed.

Omics technology applied majorly for accurate understanding of normal physiological processes and gaining knowledge related to disease processes which involves screening, diagnosis, and prognosis that provides an understanding of the etiology of diseases.

**Keywords** Omics · Systems biology · R language

P. Narad
Amity Institute of Biotechnology, Amity University, Noida, Uttar Pradesh, India
e-mail: pnarad@amity.edu

S. V. Kirthanashri (✉)
Amity Institute of Molecular Medicine & Stem Cell Research, Amity University,
Noida, Uttar Pradesh, India
e-mail: svkirthanashri@amity.edu

© Springer Nature Singapore Pte Ltd. 2018
P. Arivaradarajan, G. Misra (eds.), *Omics Approaches, Technologies And Applications*, https://doi.org/10.1007/978-981-13-2925-8_1

## 1.1 Background

In biology the suffix -omics refers to huge biological molecules; the broad analysis of large biological molecules was needed to be studied in detail as the conclusion of human genome project (HGP) in 2001. The HGP revealed that the human genome contained lesser number of genes and biological process were regulated not particularly on DNA sequence but involved various other processes, and with this evolved the new branch of study termed the omics (Hood and Rowen 2013).

This technology deciphered the cell, tissue, and organism in a holistic way around central dogma for the detection of genes (genomics), mRNA (transcriptomics), proteins (proteomics), and metabolites (metabolomics) in the samples (specific biological component). Since the technology is non-biased, they are also referred to as high-dimensional biology, while the integration of these is the systems biology. Following the discovery of DNA structure by Watson-Crick in 1953, a series of inventions and discoveries followed. The development of PCR by Kary Mullis opened all possible channels in molecular biology research. The progress in *Omics* started from the development of genomics further followed by transcriptomics and finally the proteomics, and the term was coined in 1994 by Marc Wilkins. This was possible because of advanced development in techniques like high-resolution two-dimensional electrophoresis. The cascade of events in *Omics* is depicted in Fig. 1.1. The advantage of the omics study is that they reveal specific results that promote understanding. As the omics technology is of immense potential, they have been explored in various branches of medical and health science. This technology can help to understand the etiology of disease condition through the process of screening, diagnosis, and prognosis and also for the biomarker discovery to be made easy as they involve simultaneous investigation of multiple molecules (Poisot et al. 2013). Further *Omics* is of great use in drug discovery and toxicity assessment. Pharmacogenomics deals with the connection of genomics and pharmacology to examine the role of inheritance in individual variation in drug response utilized to individualize and optimize drug therapy. They help in the field of oncology to evaluate rigorous systemic toxicity and unpredictable efficacies that are

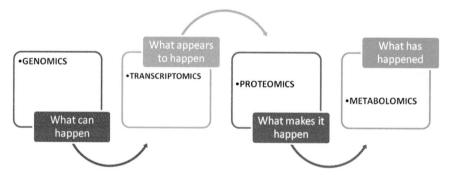

**Fig. 1.1** Cascade of *Omics* development

hallmarks of cancer therapies. These technologies are helpful in selecting novel targets for the treatment including conditions like cancer, cardiovascular disease, and obesity. In the future, systems biology promises to develop new approaches that will be predictive, preventive, and personalized (Sagner et al. 2017). Research in the field of obstetrics and gynecology is currently taking advantage of these possibilities which can be used to solve the problems related to fertility. This review aims to provide a complete overview of various omics technologies available.

## 1.2   Overview of Omics

The omics technology can be classified into various types depending on their function. Figure 1.2 highlights a few of various omics technologies that are presented in detail in the following chapters.

### 1.2.1   Genomics

This refers to the interdisciplinary study based on evaluating the structure and function and mapping of the genomes. In short this is the study of a set of genes, the inheritance substance. The term genomics was coined by Tom Roderick in 1986 on mapping the human gene. The possible and highly researched areas under genomics include the functional genomics, metagenomics, and epigenomics (Feinberg 2010).

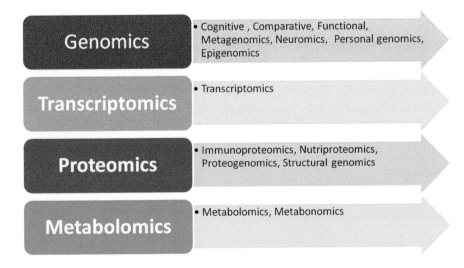

**Fig. 1.2**  Various omics technologies

## 1.2.2   Transcriptomics

The presence of mRNA in the sample reflects the abundance level of the corresponding gene. Gene expression involves the detection and classification of mRNA mixture in a specific sample. The goal of gene expression profiling is to differentiate the mRNA mixtures from different samples. Contrary to genotyping, gene expression categorizes the level of gene expression. The variation of the transcriptome can be seen over time between cell types and change according to environmental conditions (Hubank 2004).

## 1.2.3   Proteomics

The proteome refers to the total proteins expressed by a cell, tissue, or organism. The proteome is highly variable over time, shows species variation, and alters depending on environmental conditions. Proteomics is defined as the study that provides insights to protein functions in biological systems. Due to the variability and abundance of proteins in biological samples, there is a requirement to develop technologies to detect a wide range of proteins in samples of different origins. Currently exploited proteomic technologies are mass spectrometry (MS) and protein microarrays using capturing agents such as antibodies. However, the high dynamic range (abundance and concentration) of proteins complicates this type of proteomic analysis (Chandramouli and Qian 2009).

## 1.2.4   Metabolomics

The small molecules (e.g., lipids or vitamins) referred to as metabolites constitute the metabolome. The interaction between genetic, environmental, lifestyle, and other factors results in metabolic phenotypes. Interaction of metabolome with other biological macromolecules in the cell results in metabolic pathways. The metabolic profiles of biological sample represent the metabolomics which are changeable and time dependent and had a wide range of chemical structures (Bino et al. 2004).

## 1.3   Overview of Systems Biology

### 1.3.1   Systems Biology

The genome is the total DNA of a cell in the organism. The human genome contains about 3.2 billion bases with 30,000–40,000 protein-coding genes. The microarray technique enables quick analysis of the genes and also helps in examining the differences in the DNA sequences and expression of genes, which help to analyze the chromosomal abnormalities. Variation in DNA sequence leads to single nucleotide polymorphism (SNP), which plays an important role in pharmacogenomics to explore individual patient responses to drugs. The total mRNA in the cell or organism is called the transcriptome, and they are the template for protein synthesis and are known as translation. The transcriptome reflects the genes that are actively expressed at any given moment. The advent of microarray techniques has led to the advancement of the genomics and transcriptomics. Microarrays measure changes only in mRNA that makes it complex for data interpretation. Most of the mechanistic and mathematical models are used in systems biology approach for data interpretation. Due to the large number of parameters, variables, and constraints in cellular networks, numerical and computational techniques are often used (Likić et al. 2010).

## 1.4 Techniques Involved in Systems Biology

### 1.4.1 Analytical Techniques

Reproducibility of the samples is the major concern for omics experiments. Expression profiling is one of the dominant modes of DNA microarray. Complementary DNA (cDNA)/oligonucleotide is the probe to estimate the amount of mRNA in gene expression microarray. The probe gets amplified by polymerase chain reaction (PCR) and immobilized on a solid support (glass slide) by spotting them. Extraction of RNA from the sample is carried out followed by reverse transcription along with the addition of fluorescent dyes where cDNA are generated which are hybridized in the microarray slide. The chips which are the microarray glass slides are scanned by ultraviolet laser to detect the fluorescent signal produced by each gene to carry out image analysis. Samples for analysis in metabolomic experiments require fractionation (chromatography or electrophoresis) utilizing various chemical/physical properties of molecules that fasten the separation of the metabolites in liquid or gas phase. The commonly used tool for analysis of the metabolite is the mass spectrometry. The analytical techniques had their own limitations and advantages in terms of instrument sensitivity, resolution, mass accuracy, and dynamic range, while various techniques are researched to analyze the entire proteome or metabolome. For instance, the proteomics study involves characterization of proteins using electrospray ionization (ESI), matrix-assisted laser desorption/ionization (MALDI), and surface-enhanced laser desorption/ionization (SELDI) though reproducibility, accuracy, and mass range are always a limitation. The use of fluorescent tags in gel-based techniques like differential image gel electrophoresis (DIGE) and isotope-coded affinity tag (ICAT) labeling is employed with mass spectrometry to achieve better resolution. The nuclear magnetic resonance (NMR) spectroscopy and infrared spectroscopy have been routinely used for metabolite identification (van der Greef et al. 2004). Thus each technique has its unique and meritorious way of detection compared to the other techniques; it all depends also on the sample and the objective of the experiment.

### 1.4.2 Data Analysis

The analysis techniques generate huge data that mandates sophisticated software (bioinformatics and statistics). The results of the genomics and transcriptomics microarray are often huge and complicated that often conclude in false-positive results, if not accurately analyzed. Proteomics theoretical database is often matched with the experimental analysis to enable protein identification and/or quantification, while in metabolomics, raw data processing is carried out to generate meaningful and interpretable data. Thus, the prime aim of the data analysis is to represent the data in readable/understandable format which can be used to generate further hypotheses for testing with no false-positive results (van der Greef et al. 2004).

In the following text, we will discuss few of these packages using R language and their utility for analysis of "omics" data.

### 1.4.3 R Language in Omics Analysis

R is a statistical language which is fully featured and equipped with several packages useful for the "omics" and other life sciences research. It has an interactive and user-friendly interface where one can make plenty of debugging. The use of the language is coherent, and there is an extensive documentation available on the Internet to perform the data analysis. Integration to the Bioconductor platform has extended the ability of performing analysis and an easy approach for high-throughput "omics" data analysis. Within the last decades, huge amount of data has been generated through various sophisticated techniques of genomics/proteomics and metabolomics. There has been an array of new technologies in the past which have made new discoveries and research easier. It is a common practice to analyze each of the "omics" data like proteomics, genomics, and transcriptomics through statistical approaches like t-test and ANOVA. The task at hand is to make sense of the sea of data; else data generation is of no use. Toward this, R and Bioconductor platforms together provide packages for the interpretation of high-throughput data generated from "omics." There are numerous data analysis packages which offer great features to the person working on these samples. These include the packages which are computationally highly efficient for the purpose of handling large sample data; secondly these packages are able to perform reduction of the dimension by creating smaller spaces and analyzing the data; thirdly they are helpful in providing better insights to the biological system under observation. When we talk about the integrative approach for systems biology, analyzing both the datasets together is required for the understanding of the different levels of "omics." For instance, now it is clear that any integration would need inputs

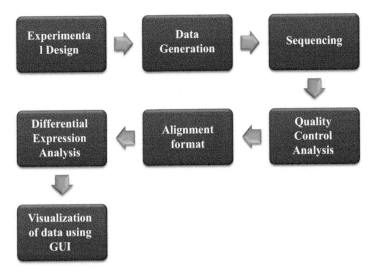

**Fig. 1.3** General workflow for "omics" data analysis

from all branches of "omics" like transcriptomics, genomics, metabolomics, and proteomics in order to understand the biological processes in a comprehensive manner.
Figure 1.3 provides a general workflow of an omics data analysis.

R and Bioconductor play an important role in extracting useful information from
large-scale high-throughput "omics" data. R console is an interactive and user-
friendly coherent language for data analysis. What makes R not the same as other
programming dialects is its GUI for fast and simple transfer of information.
Bioconductor (www.bioconductor.org) contains in silico software packages for
interpretation of "omics" data which is generated from a number of experiments
like microarray, SAGE, MS, and MS-MS. The packages at the Bioconductor platform can be split into three branches like the Annotation Data, the Experiment Data,
and the Software. Some of the important packages are listed in Table 1.1.

### 1.4.4  cpma

This package performs phenotype analysis. Numeric values from the data are treated
as the input.

### 1.4.5  mlm

This package is useful for fitting multiple linear models together. The argument
consists of a formula, which consists of the description of the models, and data,
which consists of the variable of the model.

**Table 1.1** List of the packages for quality control and analysis of gene/protein microarray data

| S. no. | Name of the package | Description |
|--------|---------------------|-------------|
| 1. | Affy package | Oligonucleotide array analysis |
| 2. | affylmGUI | GUI for analysis of one-color affymetrix data |
| 3. | ExpressionView | Visualization of possibly overlapping biclusters |
| 4. | annmap | Genome annotation and visualization package |
| 5. | DEGseq | Differential expressed gene analysis from RNA-Seq data |
| 6. | Dexus | Differentially expressed genes in RNA-Seq |
| 7. | MiChip | Differentially expressed data of miRNA for multiple species |

### 1.4.6   mixOmics

This package describes a multiple variable-based system for the "omics"-based data and its analysis of information examination to the scientist who wants to give a few appealing properties. Mostly, the package is computationally productive to deal with huge informational collections, where the quantity of sample sets is significantly bigger than the quantity of tests. Furthermore, the package performs measurement lessening by anticipating the information into a littler subspace while catching and featuring the biggest feature selection from a variety of information, bringing about great perception of the natural framework under investigation. Finally, the information appropriation makes it profoundly adaptable to answer topical inquiries over various science-related fields (Rohart et al. 2017). mixOmics multivariate strategies have been effectively connected to factually coordinate informational collections created from contrast sciences ranging from the field of "omics" comprising of transcriptomics, proteomics, and metabolomics.

### 1.4.7   integrOmics

integrOmics productively performs integrative investigations of two kinds of "omics" factors that are estimated on similar examples. It incorporates a regularized form of standard connection investigation to illuminate relationships between two datasets and a scanty rendition of incomplete slightest squares (PLS) relapse that incorporates synchronous variable choice in both datasets. The helpfulness of the two methodologies has been shown already and effectively connected in different integrative examinations (Lê Cao et al. 2009).

### 1.4.8   supraHex

supraHex is an R package for preprocessing, normalizing, and imagining omics information. This console package devises a supra-hexagonal manual to process the information, and it gives a versatile functionality for after-dissection of the guide and, all the more imperatively, takes into consideration overlaying extra information for multilayer omics information examinations. The application of supraHex was exhibited through its ability to apply DNA replication timing data, and it performs the same level of grouping and provides a real-time picture of the natural process. The scientist also observed that CpG overlaying to the replication map resulted in demonstrating the ability of supraHex to establish connections between CpG thickness and late replication data. Being part of the Bioconductor venture, supraHex is useful in making available to a wide network basically what might somehow or another be an unpredictable structure for the ultrafast comprehension of any forbidden omics information, both deductively and aesthetically (Fang and Gough 2014).

## 1.4.9 OmicCircos

OmicCircos is an R programming bundle used to create great roundabout plots for envisioning genomic varieties, including change designs, duplicate number varieties (CNVs), articulation examples, and also methylation designs. This method can be used to generate scatterplots for the abovementioned examples. Using the factual and realistic capacities in an R/Bioconductor condition, OmicCircos performs measurable examinations and presentations that come about utilizing bunch, boxplot, histogram, and heatmap groups. Moreover, OmicCircos offers various one of a kind capacity, including free track drawing for simple adjustment and combination, zoom capacities, connect polygons, and position-autonomous heatmaps supporting small representation (Hu et al. 2014).

## References

Bino, R. J., Hall, R. D., Fiehn, O., Kopka, J., Saito, K., Draper, J., Nikolau, B. J., Mendes, P., Roessner-Tunali, U., Beale, M. H., & Trethewey, R. N. (2004, September 1). Potential of metabolomics as a functional genomics tool. *Trends in Plant Science, 9*(9), 418–425.

Chandramouli, K., & Qian, P. Y. (2009). Proteomics: Challenges, techniques and possibilities to overcome biological sample complexity. *Human Genomics and Proteomics: HGP, 2009*, 1.

Fang, H., & Gough, J. (2014, January 3). supraHex: An R/Bioconductor package for tabular omics data analysis using a supra-hexagonal map. *Biochemical and Biophysical Research Communications, 443*(1), 285–289.

Feinberg, A. P. (2010, October 13). Epigenomics reveals a functional genome anatomy and a new approach to common disease. *Nature Biotechnology, 28*(10), 1049.

Hood, L., & Rowen, L. (2013, September). The human genome project: Big science transforms biology and medicine. *Genome Medicine, 5*(9), 79.

Hu, Y., Yan, C., Hsu, C. H., Chen, Q. R., Niu, K., Komatsoulis, G. A., & Meerzaman, D. (2014, January). OmicCircos: A simple-to-use R package for the circular visualization of multidimensional omics data. *Cancer Informatics, 13*, CIN-S13495.

Hubank, M. (2004, March 1). Gene expression profiling and its application in studies of haematological malignancy. *British Journal of Haematology, 124*(5), 577–594.

Lê Cao, K. A., González, I., & Déjean, S. (2009, August 25). integrOmics: An R package to unravel relationships between two omics datasets. *Bioinformatics, 25*(21), 2855–2856.

Likić, V. A., McConville, M. J., Lithgow, T., & Bacic, A. (2010). Systems biology: The next frontier for bioinformatics. *Advances in Bioinformatics, 2010*, 1.

Poisot, T., Péquin, B., & Gravel, D. (2013, April 1). High-throughput sequencing: A roadmap toward community ecology. *Ecology and Evolution, 3*(4), 1125–1139.

Rohart, F., Gautier, B., Singh, A., & Le Cao, K. A. (2017, November 3). mixOmics: An R package for 'omics feature selection and multiple data integration. *PLoS Computational Biology, 13*(11), e1005752.

Sagner, M., McNeil, A., Puska, P., Auffray, C., Price, N. D., Hood, L., Lavie, C. J., Han, Z. G., Chen, Z., Brahmachari, S. K., & McEwen, B. S. (2017, March 1). The P4 health spectrum— a predictive, preventive, personalized and participatory continuum for promoting healthspan. *Progress in Cardiovascular Diseases, 59*(5), 506–521.

van der Greef, J., Stroobant, P., & van der Heijden, R. (2004, October 1). The role of analytical sciences in medical systems biology. *Current Opinion in Chemical Biology, 8*(5), 559–565.

# Chapter 2
# Genomics

**Desh Deepak Singh and Manali Datta**

**Abstract** Genomics refers to the study of function, structure, and interactions of the genome, and it is one of the most rapidly developing scientific areas. An organism's complete set of DNA, including both protein-coding and noncoding genes, constitutes the genome. The completion of the Human Genome Project in 2003 laid a foundation for in-depth study of genomics and led to the beginning of the "genomics era." Next-generation sequencing including exome and DNA sequencing has provided a plethora of means by which we can dissect the genome at structural and functional levels. During the last decade, developments and advances in the field of genomics have led to a better understanding of human genome architecture, discovery of disease-associated genetic variants, and development of newer diagnostic methods in the field of clinical genomics. The Encyclopedia of DNA Elements (ENCODE) Project in 2010 established yet another landmark for the genomics era. The ENCODE Project characterized and annotated the functional elements hidden within the human genome's 3.2 billion bases with the aid of next-generation sequencing technologies, chromosomal conformation capture techniques, and epigenomic methods. It resolved the widespread myth about junk DNA being nonfunctional and provided evidence that the DNA between protein-coding genes consists of myriad elements (such as enhancers, silencers, and insulators) that regulate gene expression by switching transcription on or off, or by regulating messenger RNA turnover and consequently affecting translational efficiency.

This chapter provides readers with an up-to-date and coherent view of human genome architecture and also provides information about different milestones in the genomics era and new technological advances in the field.

**Keywords** Human Genome Project · ENCODE · Sequencing · Comparative genomics · Functional genomics · Clinical genomics

D. D. Singh · M. Datta (✉)
Amity Institute of Biotechnology, Amity University Rajasthan, Jaipur, Rajasthan, India
e-mail: ddsingh@jpr.amity.edu; mdatta@jpr.amity.edu

© Springer Nature Singapore Pte Ltd. 2018
P. Arivaradarajan, G. Misra (eds.), *Omics Approaches, Technologies And Applications*, https://doi.org/10.1007/978-981-13-2925-8_2

## 2.1  Introduction

The word "genome" is a conjunction of the words "gene" and "chromosome." The genome is the complete set of hereditary information for each living entity and is needed for its development and functioning. The study of genomics enables exploration of issues raised by the genomic foundation. In the 1990s, the entire genome of *Haemophilus influenzae*, a free-living organism, was sequenced, and this is considered a significant contribution to the field of genomics. The study of genomics collectively characterizes, quantifies expression and its associated regulatory network. In short, such study facilitates analysis of transcriptomic, proteomic, and epigenomic data in relation to the biological systems in prokaryotes, eukaryotes, and humans. Till date, the genomes of 6070 eukaryotes, 145,357 prokaryotes, 17,614 viruses, 12,924 plasmids, and 11,732 organelles have been sequenced and are easily accessible in the public domain in the National Center for Biotechnology Information (NCBI) genome database. This chapter describes the journey from the discovery of the human genome to application of sequencing technologies in genomics research.

## 2.2  The Human Genome Project

The Human Genome Project (HGP) create a landmark in genomic research and development. It was all initiated with the detection of the double-helical DNA structure by James Watson and Francis Crick. The possibility that the entirety of the coded information may lie in this simple coiled structure of DNA was intriguing, and more and more techniques were discovered that could make the decoding easier. The first formal proposal to study the HGP was initiated in 1984 by the Department of Energy (DOE) and the National Institutes of Health (NIH) in the USA.

After that, approvals were given in 1984 by the National Academy of Sciences (NAS) and further adapted for 5 years jointly by the NIH. The HGP was initiated in 1990 by the International Human Genome Sequencing Consortium (IGHSC)—a consortium of 20 research centers (Fig. 2.1)—with goals to make physical and genetic maps of the human genome.

The IGHSC followed hierarchical shotgun sequencing (HSS) methods and shotgun technology fragments the DNA into smaller lengths, followed by sequencing of each of the fragments. Following this, the fragments are overlapped to recreate the genome.

In 1996, the "Bermuda principles" were laid down, which agreed that any data discovered would be put in the public domain within 24 h of discovery. The Bermuda Principle was initiated to maximize the utilization of the data for research and development. Two early goals of the HGP included the preparation of genetic and physical maps of the human and mouse genomes with improved DNA sequencing technologies sequencing of smaller genomes (Yeast and worms) were taken as test template, with projected aim of technology transfer. The project took on impetus

**Members of International Human Genome Sequencing Consortium**
- ✓ The Whitehead Institute/MIT Center for Genome Research, Cambridge, Mass., U.S.
- ✓ The Wellcome Trust Sanger Institute, The Wellcome Trust Genome Campus, Hinxton, Cambridgeshire, U. K.
- ✓ Washington University School of Medicine Genome Sequencing Center, St. Louis, Mo., U.S.
- ✓ United States DOE Joint Genome Institute, Walnut Creek, Calif., U.S.
- ✓ Baylor College of Medicine Human Genome Sequencing Center, Department of Molecular and Human Genetics, Houston, Tex., U.S.
- ✓ RIKEN Genomic Sciences Center, Yokohama, Japan
- ✓ Genoscope and CNRS UMR-8030, Evry, France
- ✓ GTC Sequencing Center, Genome Therapeutics Corporation, Waltham, Mass., USA
- ✓ Department of Genome Analysis, Institute of Molecular Biotechnology, Jena, Germany
- ✓ Beijing Genomics Institute/Human Genome Center, Institute of Genetics, Chinese Academy of Sciences, Beijing, China
- ✓ Multimegabase Sequencing Center, The Institute for Systems Biology, Seattle, Wash.
- ✓ Stanford Genome Technology Center, Stanford, Calif., U.S.
- ✓ Stanford Human Genome Center and Department of Genetics, Stanford University School of Medicine, Stanford, Calif., U.S.
- ✓ University of Washington Genome Center, Seattle, Wash., U.S.
- ✓ Department of Molecular Biology, Keio University School of Medicine, Tokyo, Japan
- ✓ University of Texas Southwestern Medical Center at Dallas, Dallas, Tex., U.S.
- ✓ University of Oklahoma's Advanced Center for Genome Technology, Dept. of Chemistry and Biochemistry, University of Oklahoma, Norman, Okla., U.S.
- ✓ Max Planck Institute for Molecular Genetics, Berlin, Germany
- ✓ Cold Spring Harbor Laboratory, Lita Annenberg Hazen Genome Center, Cold Spring Harbor, N.Y., U.S.
- ✓ GBF - German Research Centre for Biotechnology, Braunschweig, Germany

**Fig. 2.1** International Human Genome Sequencing Consortium (IHGSC) members

when Dr. Craig Venter—the founder of Celera Genomics, a highly competitive enterprise—joined the race to sequence the human genome. This led to discovery of cutting-edge technologies with collaborative and competitive sharing of data, enabling improvements in data generation and collection.

Celera Genomics used Applied Biosystems automatic sequencing methodology to map the sequences and positions of expressed sequence tags (ESTs). In 1999, a sequence for human chromosome 22 (Chr 22) was reported, with the first draft of the HGP being detailed in 2001. Chr 22 was particularly chosen because of its small size and its association with various diseases. Three billion letters of the *Homo sapiens* genome were finally shared in the final draft of the publicly-funded HGP in 2003. This draft, although relatively complete, had some major gaps such as incomplete annotations and discontinuous sequences. A more updated version was finally supplemented in 2006, which gave the complete sequence of Chr 1, the largest chromosome of the human genome, containing 8% of the genetic material. Seven years later, the genome of *Homo neanderthalensis* was made public.

The first catalog, known as the "Index Marker Catalog," detailed the complete mapping with 10–15 cM resolution. The human genome was basically found to be 98.6% junk DNA. The HGP has enabled discovery of 1800 disease-causing genes, with the first one being found to be associated with Parkinson's disease and, surprisingly, 850 sites being concerned with common diseases. Most of the hot spots were found to be in the vicinity of the flanking sequences rather than within them. More than thousand genetic tests and three hundred biotechnology baased products have resulted from clinical trials based on HGP data.

During the HGP formulation, three major organizations—the Human Genome Organization (HUGO), the European Commission (EC), and the United Nations Educational, Scientific and Cultural Organization (UNESCO)—were involved in dissemination of knowledge, creating a framework for process functioning and assessing the ethical and sociolegal aspects of data sharing (Chial 2008; Cavalli-Sforza 2005).

## 2.3   Mapping of the Human Genome

Human genome mapping consists of different techniques involved in assigning a particular location of gene on a chromosome and assessing the relative distance between genes on chromosomes. The publicly funded consortium utilized HSS, in which the genome is broken into many fragments and each fragment is cloned in a bacterial artificial chromosome (BAC), using a combination of restriction enzymes and ligases. In silico assembling of the sequenced fragments results in recreation of the whole genome. This technique was utilized by the consortium to generate the first draft of the HGP with 23 billion base pairs (bp).

In contrast, Celera Genomics used a whole-genome assembly (WGA) and a compartmentalized shotgun assembly (CSA) for genome mapping. WGA follows a similar protocol to HSS strategies, whereas CSA subdivides the human genome into segments, followed by use of a shotgun sequencing method on the segments separately. In the following sections we discuss the various mapping techniques in detail.

### 2.3.1   Hierarchical Shotgun Sequencing

The HSS mechanism involves a combination of a classical strategy and a whole-genome sequencing strategy. The large genomic target is fragmented and a library is created. There are various vectors such as phages (30 kb), cosmids (50 kb), BAC (100–300 kb), and yeast artificial chromosomes (500–1000 kb) The HGP used BAC vectors for creating the library. These vectors possess the capability to replicate in bacterial hosts and hence provide an easy step for clone amplification. Generation of physical maps involves identification, as well as determination, of landmark loci within the BACs. A physical map of each insert was created using classical sequencing associated with DNA fingerprinting and fluorescence in situ hybridization (FISH) data. The locations of the contiguous DNA sequences were established by screening the sequence-tagged sites (STSs) and restriction sites. Each STS consists of a polynucleotide sequence that can be specifically detected either by a polymerase chain reaction (PCR) assay employing two oligo-deoxynucleotide primers or by hybridization experiments. Hence any two clones possessing the same STS may indicate an overlapping sequence. Similarly, each clone was digested with a set of 3–5 restriction enzymes and the profile for each was generated on agarose gel.

Two clones giving same banding pattern might contain overlapping sequences from the genome. On the basis of the data from the fingerprinting pattern, overlapping clones were identified. Subsequently, tiling was performed to involve the minimum number of BACs to cover the whole genome. The sequences thus generated are assembled into ordered arrays resulting in long-range physical maps. The contigs of the library are thus sequenced from either end using fluorescent chemical labels (Waterson et al. 2005).

## 2.3.2  Whole-Genome Sequencing

Instead of relying on a BAC-based end sequencing method, the WGA method relied on linkage maps and markers such as long interspersed nucleotide elements (LINE) and minisatellites for recreating the genome assembly. In WGA, the genome is sheared into fragments of 2 kb, 10 kb, and 50 kb, and unique sequences are identified by DNA sequencing. Any genome tends to contain multiple copies of nonfunctional repeat sequences, LINE, and minisatellites. Computational methods based on sophisticated algorithms were used to identify overlapping DNA sequences on the basis of the presence of repeat elements at either end; this method was aptly named mate pairing. Contigs with mate pairs are parsed together and addressed as uniquely assemblable contigs (unitigs). A 30-fold reduction in "sheared pieces" and a 100-fold reduction in overlaps were observed with unitig identification. Unitigs were subsequently reassembled as scaffolds on the basis of appearance of the same repeat elements and consensus sequence comparison, and the genome was reconstructed from the sets of scaffolds.

## 2.3.3  Haplotype Genome Sequencing

In eukaryotes, the genome tends to possess different levels of ploidy. In humans, two sets of chromosomes are present and hence are known as diploids. With the data generated from HSS and WGS, it is extremely difficult to identify the haplotype of the given individual; hence, haplotype genome sequencing has become a requirement. Haplotype information is crucial for understanding linkage analysis, genomics studies, and clinical genetics for diagnosis and treatment of patients. Haplotype mapping (HM) can be broadly classified into sparse and dense HM (DHM). In DHM, genomic DNA can be extracted from a sample containing a total population of cells.

Genomic DNA is gently extracted from cells. The extracted sample tends to be a mixture of both haplotypes from the whole genome. High molecular weight (HMW) DNA is categorized by size on the basis of the gel electrophoresis profile. Enrichment of fragments with sizes ranging from 10 to 100 kb is done by compartmentalization. Pooled DNA is subsequently cloned in fosmid vector, packaged in phage, and used to transduce bacteria for library propagation and outgrowth, then the library is erratically diluted to a large number of reaction blocks, so that respectively each block has zero to one copy of any region of the genome containing a single haplotype at any locus. The library is sequenced and indexed to generate information about the presence of heterozygous variants.

An additional adapted procedure of HM is contiguity preserving transposition (CPT)Seq that utilizes Tn5 transposase. This barcoded enzyme binds strongly to HMW DNA after "tagmentation" through indexed DNA connectors. Alleles co-occurring in the similar HMW haplotype are physically preserved at this stage; afterward sequence of transposition and thinning ladders, a protein denaturation step leading to the removal of Tn5 transposase and formation of pronounced and indexed disjointed templates. The templates are subsequently amplified by PCR to introduce a second index. As this intensification step functions on ~200-bp fragments rather than on HMW DNA the subsequent public library consistency is better as comparable to other in vitro methods. In each method, all libraries are sequenced, employing the properties of conventional shotgun sequencing followed by variant calling.

Fluorophore-labeled metaphase chromosomes are separated from single nucleus based on fluorescence-activated cell sorting (FACS) into different reaction chambers. Every chamber containing one chromosome therefore contains a single haplotype. Discrete chromosomes thus separated are immobilized on a microscope slide followed by targeted genotyping by PCR and single-base extensions. Sequenced reads are aligned to the genome and matched up to heterozygous variants orthogonally. Since the same haplotypes produce the sequenced fragments, heterozygous sites falling within "islands" of coverage are connected to form chromosome-length haplotypes (Snyder et al. 2015).

## 2.4  DNA Sequencing

Sequencing of DNA involves determining the order of the four biochemical structure chunks—called "bases"—that make up the DNA molecule. With the advent of knowledge and the requirement for ultra-low-cost sequencing (ULCS) methodologies, three main generations of sequencing technology have arisen. These encompass any one of five technologies namely microelectrophoretic methods, cyclic-array sequencing on amplified molecules, 'sequencing by hybridization', cyclic-array sequencing on single molecules and non-cyclical, single-molecule, and real-time methods (Fig. 2.2).

### 2.4.1  First-Generation Sequencing

Sequencing techniques have advanced by leaps and bounds. The first-generation sequencing methods are basically modified methods of Sanger's sequencing with fluorophores being used for detection by fluorescence based techniques.

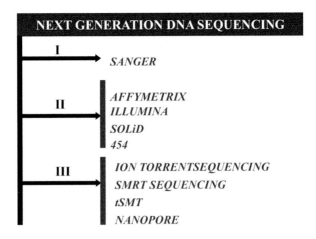

**Fig. 2.2** Next-generation sequencing techniques, highlighting sequencing methodologies of the first, second, and third generations

#### 2.4.1.1   Sanger Method

One of the foremost mechanisms of DNA sequencing is the dideoxy chain termination method. It was discovered by Sanger so it is also known as the Sanger method. This is a micro electrophoretic method that replicates the template strand of DNA to be sequenced and impede the replication procedure by insertion of dideoxy nucleotide bases. Knowledge of nucleotide end sequences of the target is a prerequisite for this technique. A replica of the respective template is separated into four groups, and each group is applied for a diverse duplication response. Copies of a standard primer and DNA polymerase I are used in all four batches. To produce fragments that terminate at A, ddATP is added to the reaction mixture in batch 1 along with dATP, dTTP, dCTP, and dGTP, average primer, and DNA polymerase I. A maximum of 384 capillary mechanized parallel reactions can be run for Sanger sequencing. An advancement in Sanger sequencing, known as dye terminator sequencing, comprises four ddNTP chain terminators labeled with fluorescent dyes, each of which emits light at a different wavelength. As the fragments of different lengths are being resolved in the capillary, an excitation laser shines through it and enables detection of bands emitting different colors owing to their fluorescent tags. As the respectively colored band is identified, a signal is fashioned, administered by the sequencer, and obtainable as a peak on a graph, whereby the respectively peaks characterize a diverse base.

### 2.4.2   Second-Generation Sequencing

Second-generation sequencing (SGS) is the current industry standard and is used to read lengths that are roughly 100–500 base pairs at most. The number of sequences that all cover the same region is known as "coverage" and is essential as possible errors may happen in the sequencing preparation or in amplification, or as a result of the sample being heterogeneous, (Difference in ploidy of genome). SGS techniques rely on emulsion PCR and bridge PCR for amplification of signals. We will briefly discuss these two PCR modifications.

### 2.4.2.1  Emulsion PCR

Emulsion PCR (EmPCR) works on the principle of dilution and compartmentalization of the sample. Genomic DNA is fragmented either by sonication or by nebulization, and adaptors are attached to the $3'$ and $5'$ ends. Streptavidin-conjugated beads immobilized with the same adaptors are synthesized and placed with the modified fragmented DNA. Each conjuction procedure generates one bead with a specific type of dsDNA acting as a multiple microreactor. A denaturation step separates the library fragment into two distinct elements, with subsequent annealing of the complementary element to the bead. The annealed DNA is amplified by polymerase starting from the bead towards the primer site. The process is then repeated over 30–60 cycles leading to clusters of DNA. The amplified DNA is obtained by breaking the emulsion, followed by bead enrichment.

### 2.4.2.2  Bridge PCR

Genomic DNA for bridge PCR is processed in a similar way with attachment of adaptors. A flow cell is designed whereby the surface is thickly coated with adaptors that have ability to bind the connecters conjugated to the DNA library fragments. The amplified sample is denatured and passed through this flow cell, tends to get attached to the surface of the flowcell at random. On addition of nucleotides and enzymes, the free ends of the single strands of DNA forms bridged structures, where it is exposed to reagents for polymerase based extension. The linked structure enables formation of clonal clusters of localized identical dsDNA.

### 2.4.2.3  454 Sequencing

A pioneer technique developed by the 454 Corporation, this was the first commercially available next-generation sequencer. 454 sequencing works on the 'sequencing by synthesis' principle, whereby complementary strand synthesis is mediated by polymerase enzyme. Pyrosequencing relies on the release of pyrophosphate with addition of nucleotide to each DNA chain. Release of pyrophosphate ensures a sequential reaction of adenosine phosphosulphonate in the presence of adenosine triphosphate (ATP) sulfurylase and subsequent conversion to ATP. The ATP reacts with the luciferin present as a constituent of the reaction mixture in luciferase-catalyzed conversion and emits pulses of light (oxyluciferin). The relative intensity of light is proportional to the amount of base added (i.e., a peak of twice the intensity indicates two identical bases have been added in succession).

#### 2.4.2.4 Sequencing by Ligation (SOLiD)

This method was developed in 2007 and involves genomic library construction and ligation followed by sequencing. EmPCR is utilized to amplify the target sample followed by anchoring of agarose beads on a glass surface. Once attached, the arrangement is flooded with fluorescent-labeled oligonucleotides. The presence of complementarity between the template and the oligonucleotide allows annealing followed by ligation mediated by DNA ligase. A phosphonothioate linkage between bases enables fluorescent dye to be removed from the fragment using silver ions. This allows four different fluorescent peaks to be detected, each corresponding to different nucleotides. Removal of the fluorophore makes a 5′- become vulnerable to additional ligation.

#### 2.4.2.5 Reversible Terminator Sequencing (Illumina)

Reversible terminator sequencing, also known as sequencing by synthesis, is a part of NGS technology and was developed by Solexa in 2006. Modified nucleotides that are fluorescently labeled are used as reversible terminators. Reversible terminators can be gathered into two categories: 3′-$O$-blocked rescindable terminators and 3′-unblocked reversible terminators. Each of the four DNA bases has a diverse fluorophore attached to a nitrogenous base in addition to a 3′-$O$-azidomethyl group. The fluorescent tag is cleaved using tris(2-carboxyethyl) phosphine (TCEP), concurrently eliminating the 3′-$O$-azidomethyl group and recreating 3′-OH, and the cycle may proceed repeatedly. 3′ reversible terminators are attached to both the nucleotide and the fluorescence group, thereby acting both as part of the termination group and as a reporter. This technique varies as the 3′-position is not congested (i.e., the base has a free 3′-OH); the fluorophore remains the same for all four bases, and each modified base is flooded successively rather than at the same time. Adjustable termination uses bridge PCR, whereby addition of the nucleotide is followed by cleavage of the fluorescent terminator, generating a pulse that improves the efficiency of this stage of the procedure. The scheme practically eliminates error-prone reading and missed reads associated with strings of homopolymers. Pioneered by Illumina (HiSeq and MiSeq), second-generation sequencers can analyze a million bases at a cost of $0.02. With a high data output of 600 Gb per run, it takes 8 days to completely sequence the human genome.

#### 2.4.2.6   Affymetrix Microarray Technique

Affymetrix GeneChips are generated using solid-phase chemical synthesis. Photolithography oligoprobes are directly synthesized onto a glass wafer and may contain as many as 900,000 different variations of similar oligos. The locally synthesized probes tend to have a known location and hence hybridization designs and signal strengths can be understood in footings of gene individuality and comparative appearance heights by Affymetrix GeneChip Operating Software.

### 2.4.3   Third-Generation Sequencing

Some novel methods have subsequently been prepared using single-molecule sequencing (SMS) and single real-time sequencing, thereby reducing costs and avoiding potential biases. SMS is currently being employed by the Pacific Biosciences platform (single-molecule real-time), Oxford Nanopore Technologies, and Helicos Biosciences.

#### 2.4.3.1   Ion Torrent Semiconductor Sequencing

In ion torrent sequencing a semiconductor chip detects the hydrogen ions formed during DNA polymerization. Ion torrent sequencing is the first commercial technique not to use fluorescence and camera scanning; it translates nucleotide composition directly into digital information (0, 1) on the chip. Hence, it is a faster and less expensive as compared to previous sequencing platforms. Nucleotide incorporation into extending DNA results in release of a hydronium ion and is detected by the sequencer's pH sensor. This is instantly converted into a voltage by the semiconductor chip.

The Personal Genome Machine (PGM) sequencer, from ThermoFisher Scientific, sequentially floods the chip with one nucleotide after another. If the next nucleotide that floods the chip is not a match, no voltage charge will be recorded. The recording of the data occurs in a matter of seconds and thus it is a simple, scalable sequencing solution.

#### 2.4.3.2   Single-Molecule Real-Time Sequencing

This innovative technique was developed by PacBio and is built upon two key novelties: phospholinked nucleotide bases and zero-mode waveguides (ZMWs). A single molecule of DNA is immobilized on a ZMW and incubated with a single moiety of DNA polymerase. ZMWs allow light to illuminate only the bottom of a well in which a DNA polymerase/template complex is immobilized and creates an observation volume small enough to monitor addition of single nucleotides to the extending template. Phospholinked nucleotides permit identification of the immobilized complex as the DNA polymerase produces a completely natural DNA strand.

### 2.4.3.3   True Single-Molecule Sequencing

Sample DNA is treated with restriction enzymes and tagged with *polyA* tails. *polyT* chains bound to HeliScope flow cell plates are used to immobilize the tagged sample DNA. Labeling is performed in "quads" consisting of four cycles each, for each type of base. Fluorescent-labeled bases are added, and a laser illuminates the label, thus eliciting a signal, which can be picked up by detector. Each hybridized template is sequenced simultaneously, enabling multiple reads in a single time frame. The label is then cleaved, and the next cycle begins with a new base.

### 2.4.3.4   Nanopore Technology

Oxford Nanopore has developed a working principle in which ss-polynucleotides container is passed in a single-file through a hemolysin nanopore on a plate with millions of tiny wells with the bottom layered with DNA polymerase. The polymerase tends to grab a piece of DNA that is exposed in the well and is made to pass through a 1.5-nm nanopore. Obstruction of the pore to varying degrees by four distinct nucleotides tends to generate tiny differences in electrical conductance across the pore. As a consequence, a sequence is generated each time a piece of DNA goes through a well. The accuracy of base calling range from 60% for single events to 99.9% for 15 events (Kchouk et al. 2017).

## 2.5   Genome Annotations

Genome annotation is a procedure for of classifying the loci with assigned functionality of each gene thus identified. Genome annotation includes mapping of topographies such as protein-coding genes and their corresponding messenger RNA (mRNA), pseudogenes, transposons, repeats, noncoding RNA (NcRNA), and single-nucleotide polymorphisms (SNPs), as well as regions of resemblance to other genomes, onto genomic scaffolds. Annotation basically gives an overview as to what the species will look like through its life cycle, its mechanisms for various life processes, and its responses to the environment. There are various approaches for annotating a genome. Here, we discuss the various dimensions of annotation.

### 2.5.1 Nucleotide Annotation

This encompasses the identification of the sequence of a gene. Eukaryotic genes comprise introns and exons, which are unique and serve as markers for identification. Many computational platforms are available that annotate the start sites, termination sites, exons, introns, and exon–intron boundary for a given gene sequence. A key feature distinguishing each of these programs are sensor algorithms that identifies structural attributes, such as presence of consensus splice site junctions (GT/AG). Once the loci and sequence are characterized, the next step is naming of genes.

### 2.5.2 Naming of Genes

The software most commonly used to designate a gene name is the Basic Local Alignment Search Tool (BLAST). On the basis of the query string, this program looks through a database for sequence similarities. The most typical databases searched are GenBank and Swiss-Prot and can be either a protein or nucleotide sequence. Another database frequently referenced is the EST database, which houses the coding tags of the various sequences spread across the genome. ESTs are tags of expressed genes identified in the complementary DNA (cDNA) library. They are transcribed from functioning genes; hence, the EST database is a direct indicator of predicted coding genes. Nucleotide stretches present in the genome may be non-coding genes or marker sequences.

### 2.5.3 Annotation of Nongene DNA Sequences

RNA sequences such as ribosomal RNA and transfer RNA (tRNA), which are essential for protein translation and RNA splicing, and piwi RNA constitute this fraction of the genome. These sequences show a high degree of conservation and consequently are easily identifiable. Specific regulatory regions described by short sequences (motifs) are the key features identifying whether a gene will be expressed or not. Repetitive elements can populate genomes to larger degrees. For example, Cis-regulatory elements (CREs) are regions of non-coding DNA which control the transcription of neighboring genes. CREs constitute significant elements of genetic regulatory networks, which in turn direct morphogenesis, the growth of anatomy, and other aspects of embryonic expansion. Transposons and retrotransposons are movable genetic elements. Retrotransposon repeat sequences, which comprise long interspersed nuclear elements (LINEs) and small interspersed nuclear elements (SINEs), provide clarification for a great number of the genomic arrangements in numerous species. More than 8% of the human genome is made up of endogenous retrovirus sequences, out of which 42% fraction results from retrotransposon based events, and 3% can be recognized to be the leftovers of DNA transposons mediated events. Hence, annotation confers an identity on the other string of nucleotides arranged in a primary structure (Koonin and Galperin 2003).

## 2.6 Functional Genomics

Functional genomics is a branch of genomics where sequence data are functionally annotated and their effects are studied in generation of a cellular phenotype. It quantifies the expression levels of RNA and proteins, and allows us to comprehend the biological importance of the genes in a cell. A genome-wide approach is a preferred way of studying functional genomics, as the cross play between different moieties determines the functioning of the cell. Various techniques have now arisen to generate data for functional annotation. In the following sections, we briefly discuss some cutting-edge technologies currently in practice (Jia et al. 2017).

### 2.6.1 qPCR

Quantitative PCR (qPCR) is also known as real-time PCR. qPCR may be used to study gene expression and copy number variation, SNP genotyping, detection of rare mutations, and miRNA analysis. Real-time PCR allows dependable reliable detection and measurement of products generated during each cycle of the PCR process on the basis of detection of fluorescent probes. The TaqMan assay is one of the original options for real-time PCR. The method exploits the 5′ endonuclease activity of *Taq* DNA polymerase to cut an oligonucleotide probe throughout the PCR, thus producing a visible signal. The probes are fluorescently labeled at their 5′ end and are non-extendable at their 3′ end due to chemical modification. Specificity is conferred at three levels: via two PCR primers and the probe.

### 2.6.2 Two-Hybrid System

The yeast two-hybrid (Y2H) technique is used in recombinant yeast strains. Transcription of a discerning marker allows expression of a detailed phenotype, typically by growth on a selective medium or a change in the color of the yeast colonies (Selective marker). In colorimetric assays, for example, HIS3 enables selection of recombinant yeast on a medium lacking histidine, and LacZ to screen recombinant strains. Various yeast two-hybrid have been developed to develop screen for detection of co-factor, or identification of post-translational modification of the protein partners. Lastly, yeast n-hybrid protocols have been used to screen for novel molecular interactions such as protein–protein, DNA–protein, and RNA–protein interactions. Utilization of yeast cells has a major disadvantage: protein tends to be overtly glycosylated, hence mammalian two-hybrid systems have been designed and are being utilized. The vectors expressin fusion proteins are cotransfected with a reporter chloramphenicol acetyltransferase (CAT) vector into a mammalian cell line. The reporter plasmid contains a CAT gene under the control of five consensus Gal4 binding sites. Interaction of both fusion proteins leads to a significant increase in expression of the CAT reporter gene.

### 2.6.3 RNA-seq

RNA sequencing is the study of transcriptomic expression. It allows detection of variations of events in disease states and responses to therapeutics, different environmental stresses, and a broad range of stimuli. RNA-seq is dependent on deep-sequencing technologies whereby a population of RNA is reverse transcribed to cDNA and tagged with adaptors at one or both ends. Each molecule, with or without amplification, is formerly sequenced by either single-end sequencing or pair-end sequencing. High-throughput sequencing has been used for RNA-seq, such as Illumina, SOLiD, and 454 sequencing systems.

### 2.6.4 DNA Microarrays

In the microarray technique, DNA fragments are fixed to a substrate and then probed with a known gene sequence. An array experiment makes use of common assay templates such as microplates or membranes. The sample spots are less than 200 microns in diameter and regularly comprise thousands of spots containing probes with known sequences of DNA, cDNA, or oligonucleotides. The analysis involves binding of the probe to the target sequence, therefore determining whether expression of the target genes has occured. Instantaenously thousands of gene are detected on a one DNA chip. DNA microarrays are set up for numerous experimental goals. For monitoring expression analysis, cDNA resulting from the mRNA of identified genes is immobilized on a solid support and identified from normal and diseased tissues. Higher intensity reads are gained for a disease gene if the gene is overexpressed in the active disease. This expression design is then associated with the expression pattern of the gene responsible for the disease.

## 2.7 The ENCODE Project

The Encyclopedia of DNA Elements (ENCODE) Project was launched as a follow-up to the HGP. ENCODE was initiated in September 2003 by the National Human Genome Research Institute (NHGRI) to overcome the shortfalls of the HGP. The aim of the project was to identify functional regions as coding or noncoding. In the human genome, the topographies are mapped and consistent methods for RNA expression analysis (RNA-seq, CAGE, RNA-PET, and manual annotation), protein-coding regions (mass spectrometry), TF-binding sites (ChIP-seq and DNase-seq), chromatin structure (DNase-seq, FAIRE-seq, histone ChIP-seq and MNase-seq), and DNA methylation sites (RRBS assay) are covered (Table 2.1). In the primary stage of the project, covering 1% of the genome, the ENCODE Project annotated 60% of mammalian bases. Then, construction and preliminary examination of 1640 data sets was performed to annotate functional elements in the complete human genome (ENCODE Project Consortium 2012). As a result of ENCODE, the following observations and interpretations were made:

1. About 80.4% of the human genome is involved in at least one event in at least one cell type.
2. Ninety-five percent of the genome lies within 8 kb of a DNA–protein interacting region.
3. Eighty-six percent of the DNA segments occupied by sequence-specific transcription factors contain a strong DNA-binding motif; 2.89 million nonoverlapping DNaseI-sensitive sites (DHSs) were identified by DNase-seq in 125 cell types.

**Table 2.1** Techniques and their principles used in the Encyclopedia of DNA Elements (ENCODE) Project

| Technique | Principle |
|---|---|
| RNA-seq | Seperation of an RNA sample from sequences tracked by high-throughput sequencing analysis |
| CAGE | Capture of methylated cap at the 5′ end of RNA sequences followed by high-throughput sequencing analysis |
| RNA-PET | Simultaneous capture of RNAs at 5′ methyl cap and polyA tail followed by sequencing of a short tag from each end |
| ChIP-seq | Chromatin immunoprecipitation followed by sequencing to map DNA-binding proteins |
| DNaseI-Seq | Identify regulatory sequences with DNase I as it preferentially cleaves chromatin at specific sites followed by high throughput sequencing |
| FAIRE-seq | Exploits the difference in crosslinking efficiency between nucleosomes and sequence-specific regulatory factors |
| RRBS | Documentation and description of methylated and nonmethylated CpG stretches |

*CAGE* cap analysis gene expression, *ChIP* chromatin immunoprecipitation, *FAIRE* formaldehyde-assisted isolation of regulatory elements, *PET* paired end tag, *RRBS* reduced representation bisulfite sequencing, *seq* sequencing

4. Genome enhancers and 70,292 regions with probable promoters were identified.
5. Sixty-two percent of the genome was found to be highly enriched for histone modifications.
6. Noncoding variants in genome sequences were ascertained as ENCODE-annotated functional regions.
7. Disease-associated SNPs were found to be functional elements in the vicinity of the protein-coding regions

## 2.8 Comparative Genomics

With the advent of cutting-edge technologies, many genomes are being discovered and curated. The study of the alignment of sequences of available genomic data using computer-based tools is known as comparative genomics. Researchers have sequenced the complete genomes of numerous animals (more than 250 animal species and 50 species of birds) and plants, which includes almost 250 animal and 50 birds alone, and the list continues to grow daily. Comparative genomics enables scientists to identify genes and obtain a broader insight into the structure–function relationships of genes. Comparative in silico analysis identifies gene expression profiles, protein–protein interactions, and genetic and regulatory interactions. These in turn enable us to obtain a broad insight into the origin and evolution of cellular interactions. Such comparisons have shown that 90% of the human genome is similar to that of the mouse, while 60% conservation is observed between humans and fruit flies.

Surprisingly, regions that are highly conserved in vertebrates are prone to accelerated evolution in humans, and most frequently they constitute the regulatory regions (human accelerated regions (HARs)). Comparisons may enable identification of syntenic gradients in a species-specific manner and thus may further aid recognition of as-yet-unidentified regulatory regions in metabolic pathways. Drug targets in many infectious and metabolic diseases could be identified on the basis of comparative genomics. As a part of molecular medicine, comparison of the genomes of healthy individuals with the genetic makeup of a diseased individual may reveal clues to eliminating that disease. The ENCODE Project could methodically classify 80% of the human genome using experimental markers such as transcription and histone modification. The Model Organism Encyclopedia of DNA Elements (modENCODE) has successfully used comparative genomic techniques to understand the functionality of various human and animal genomes. As a part of modENCODE, similar patterns of gene expression and regulation could be contemplated among fly, worm, and human genomes.

Chromosome painting is one of the techniques that can be used to visually assess the similarity of two closely related species such as humans and apes. A variant of fluorescence in situ technique is based on utilization of a probe designed from a whole chromosome or from a region of a chromosome. Chromosomes to be analyzed are separated by a sorter or by microdissection, followed by amplification of the chromosome. Single type of paint is applied to individual species followed by suppression hybridization; annealing of chromosomes from two different sources may be easily identified based on their characteristic color of the chromosome paint. The fluoroscent paint acts as a tag and hence are easily distinguished. Instead of painting the whole chromosome, regions of chromosomes may also be highlighted (single G-bands) by microdissection.

A model organism possesses an idealized and simple system, which is easily accessible and may be manipulated. Comparative genomics has enabled recognition of model organisms. Some model organisms identified are the rhesus macaque (*Macaca mulatta*) (for models of human immunodeficiency virus), the chicken (for embryonic development and the role of viruses in cancer), the sea urchin (for switching of developmental genes), and the chimpanzee (for disease models), to name a few.

Genome comparison has enabled us to generate an evolutionary pattern for living organisms. Variation in the genome could facilitate quantification of divergence and convergence in different species at the molecular level. A centralized database compiling data on the complexity of genome organization, gene function, and regulatory pathways of plants has been established. Aptly named PLAZA, this platform for plant comparative genomics (http://bioinformatics.psb.ugent.be/plaza/) covers information pertaining to homologous gene families, sequence alignments, evolutionary trees, dot plots, and genomic collinearity between species. Another site, Phytozome (http://www.phytozome.net)—in addition to providing evolutionary history at the sequence, gene, and family levels—also provides annotated sequence information. Another computational platform, Compagen (http://www.compagen.org) could classify early members and the branch point of poriferans and cnidarians.

Technological advancements have enabled generation of ample data for performing comparative analysis. One of the basic approaches to obtain data for comparative genomics is performing sequence alignment. Long indels and genomic rearrangements are some major drawbacks for obtaining efficient results. Modified and advanced algorithm usage on platforms such as BLASTn and MegaBLAST (www.ncbi.nlm. nih.gov/BLAST/), GLASS (crossspeciency.lcs.mit.edu/), MUMmer (www.tigr.org/software/mummer/), PatternHunter (www.bioinformaticssolutions. com/products/ph.php), PipMaker (http://bio.cse.psu.edu/pipmaker/), and WABA (www.cse.ucsc.edu/kent/xenoAli/) has enabled comparative visualization of genomes.

The genome structure (Fig. 2.3) may be an essential criterion to perform genome comparison. Comparison of nucleotide composition has also enabled us to monitor the transfer of genes from one organism to another. There are several islands containing different (G + C) percentages that are strain specific. The study of genome signatures enables us to identify candidates for horizontal gene transfer. Exchange of fragments in conjunction with chromosomal breakage information tends to cause disruption of gene order. Hence, difference in gene order is a direct indicator of evolutionary distance between genomes (Haubold and Wiehe 2004; Wei et al. 2002).

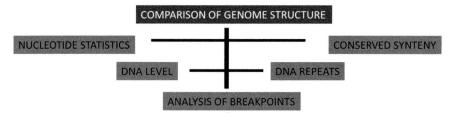

**Fig. 2.3** The genome structure may be compared at different levels

## 2.9   The Epigenome and Epigenetics

The multitude of chemical entities regulating the expression of the genome without affecting the sequence of DNA are known as the constituents of the epigenome, and the phenomenon associated with it is defined as epigenetics. These chemical components have the ability to turn expression of genes on and off for formation of functionally specialized cells and are invariably heritable. All cells essentially contain the same genome, but the cross talk of the epigenetic modifications tends to assign specific roles to cells. Lifestyle, in addition to environmental factors, tends to alter the chemical responses and thus may aggravate anomalies in expression.

Such a phenomenon has been observed in lambda phages, where alteration in the expression of regulators results in two epiphenotypes. These distinct epiphenotypes are responsible for the lysogenic and lytic phases of replication in phages. cro and cI proteins are the regulators that enable the shifting between lytic and lysogenic cycles. cI protein hinders the expression of the other, by binding to two of the three operator sequences and thus promoting its own production and hence lysogeny. Environmental fluctuations, and thus DNA damage, augment the level of Cro protein, inducing the lytic cycle of the phage. Thus, knowledge of epigenomics and epigenetics is essential to understand the healthy functioning of living organisms.

In eukaryotes, epigenetic modulations may occur via "marking" or by "regulation." There are three basic types of epigenetic modifications—DNA methylation, histone modification, and NcRNA—enabling marking and regulation of gene expression. Various technologies are available to scrutinize the prevalence of epigenetic modifications in genomes (Table 2.2).

### 2.9.1   Epigenetic Mechanisms of Regulation

Epigenetic changes are reversible, and understanding of the mechanistic basis for their efficacy may enable the creation of as-yet-unexplored therapeutics for malignancies. In the next section, we explore the mechanisms and their effects on the functioning of the gengenome (Yan et al. 2015).

**Table 2.2** Assays used for epigenetic studies

| Epigenome assay | Features |
|---|---|
| MNase-seq | Maps both histone and nonhistone proteins |
| DNase-seq | Maps cis-regulatory regions and resolution footprints for transcription factors |
| FAIRE-seq | Maps cis-regulatory regions |
| ChIP-seq | Maps DNA-binding proteins |
| ATAC-seq | Maps nucleosome positioning, chromatin accessibility, and transcription factor binding sites simultaneously |
| WGBS | Covers approximately 95% of CpGs |
| RRBS | Covers approximately 10–20% of CpGs, largely in CpG islands |
| MeDIP-seq | Covers approximately 60–90% of CpGs |
| MBD-seq | Covers 60% of CpGs |
| 27K array | Covers 27,578 CpGs |
| 450K array | Covers 482,421 CpGs |

*ATAC* assay for transposase-accessible chromatin, *ChIP* chromatin immunoprecipitation, *FAIRE* formaldehyde-assisted isolation of regulatory elements, *MBD* methyl-CpG binding domain, *MeDIP* methylated DNA immunoprecipitation, *MNase* micrococcal nuclease, *RRBS* reduced representation bisulfite sequencing, *seq* sequencing, *WGBS* whole-genome bisulfite sequencing

### 2.9.1.1  DNA Methylation

DNA methylation was discovered in 1964 by Hotchkiss and correlated with gene expression by Riggs, Holliday, and Pugh. It involves transfer of methyl from S-adenosyl methionine (SAM) to cytosine residues at position 5 especially concentrated on the cytosine guanine di-repeats known as CpG islands. Approximately 29,000 CpG sequences have been identified to date, and these have been associated with promoters that are unmethylated in normal cells, thus enabling access to transcription factors and chromatin-associated proteins for the expression of most housekeeping genes. Transfer is mediated by the family of DNA methyltransferase (DNMT). DNMT1, which was found to be associated with the S phase of the cell cycle, enables maintenance of the DNA in its heterochromatic form. Two tissue-specific methyltransferases—DNMT3a and DNMT3b—play a crucial role in embryonic differentiation and development. Methylated cytosine tends to be concentrated (70–80%) in CpGs, mainly in transposons, centromeres, and telomeres. In mammals, six methyl-CpG binding proteins (MBPs)—methylcytosine binding protein 2 (MECP2), MBD1, MBD2, MBD3, MBD4, and Kaiso—interact with nucleosome remodeling complex (NuRD) to methylate DNA, thus being a vital component of epigenetic gene regulation.

Methylation-mediated transcriptional repression involves three basic mechanisms: prevention of transcription factor binding to regulatory sequences, employment of MBPs to prevent the binding of the transcription machinery at the promoter sequence, and condensing of DNA in chromatin.

Active demethylation depends on demethylase protein complexes such as the ten–eleven translocation (TET) family of proteins (TET1, TET2, and TET3), thymine DNA glycosylase (TDG), and base excision repair (BER). DNA demethylation can be either passive (a process arising because of DNA replication without coupled de novo methylation) or active (a process involving the intermediates 5-hydroxymethylcytosine (5hmC), 5-formylcytosine (5fC), and 5-carboxycytosine (5caC)).

### 2.9.1.2  Histone Modifications

Histone is a basic constituent of the nucleosome, consisting of eight subunits of protein (two copies of H2 histones (A and B), H3, and H4) around which 147 bp of superhelical DNA is wrapped. There are distinct forms of histone modifications, each having different implications: H3K56 acetylation, H3R42 methylation, and H3K122 and H3K64 acetylation destabilize nucleosomes, whereas phosphorylation alters chromatin architecture and H1 citrullination decreases DNA interaction. Some key enzymes regulating the histone modifications are histone acetyltransferases (HATs) and deacetylases (HDACs), along with kinases, phosphatases, ubiquitin ligases and deubiquitinases, methyltransferases (HMTs), demethylases (HDMs), SUMO ligases, and proteases.

Lysine 56 in histone H3 is the entry point of DNA in a nucleosome. Hyperacetylation and hypoacetylation have been observed for the residue in the case of hypertranscribed and silent DNA stretches. This modification affects water-mediated contact between DNA and histone, resulting in compaction of the chromatin structure. Another similar observation has been made in the case of methylation of arginine 42 in H3. At the axis of the nucleosome, the binding between histone and DNA is strongest. Upon acetylation, a lysine at this juncture, K122, disrupts the stable interaction of DNA and protein. This results in unwinding of DNA from its tight binding, making it more accessible for the transcriptional apparatus. K64 of histones strategically placed near promoters of DNA releases the nucleosomes on acetylation, thus leading to increased expression. Conversely, methylation of K64 and K122 leads to tighter binding of the DNA–histone interactions, and acetylation at K56 enables the docking of chaperones to wind the DNA around the nucleosome. K56 acetylation at H4 has been found to increase the nucleosome association. Methylation on H2A (Q105) is concentrated at the recombinant DNA (rDNA) repeats and represses binding of the FACT chaperone, which in turn inhibits nucleosome reassembly.

Histone phosphorylation is limited to serine, threonine, and tyrosine residues. Phosphohistone interacting proteins belong to the 14–3–3 family. These members have been implicated in interacting with different transcriptional regulators and chromatin-modifying proteins, such as TATA-binding protein (TBP) and histone deacetylases (HDACs).

Interestingly, it has been noticed that two post-translational modifications (PTMs) tend to direct the same region of the histone protein. Two distinct models have been suggested for this occurrence. The spatially linked theory proposed by Mahadevan et al. explains that both PTMs are exclusive of each other and do not require the presence of each other for functioning. Conversely, the second hypothesis suggests that simultaneous recruitment of histone acetyltransferases and kinases is a necessity and has been experimentally proved in yeast. Another binary switch model involves phosphorylation and methylation, and is known as the phospho/methyl switch. One such switch has been observed in the Lys 9/Ser 10 region of H3. Methylation of K9 by SET DNMTs is associated with the maintenance of heterochromatin and hence gene silencing. Evidence of four such methyl/phos switches has been confirmed in the H3 histone (Tessarz and Kouzarides 2014).

### 2.9.1.3 Noncoding RNAs

NcRNAs have been observed to be dynamically involved in maintenance and expression of the epigenome and have aptly been designated as "signals" as they are expression-free activity-driven moieties. There are variants that are structurally and functionally distinctive and tend to follow "semiconservative" transmission. RNA-directed DNA methylation (RdDM) seems to be the focal epigenetic pathway in plants. It recruits a dedicated transcriptional machinery composed of two RNA polymerases—Pol IV and Pol V—targeting specific DNA sequences. Transcriptional repression mediated by RdDM is functionally evident in pathogen defense, stress responses, allelic communications, and reproduction.

In eukaryotes, an apparent role of NcRNA has also been observed during fertilization and embryo implantation, with a close association of DNMTs and polycomb proteins. Interestingly, loci expressing NcRNA are themselves prone to epigenetic modifications and thus expression. Polycomb proteins tend to mediate histone modifications and gene repression, resulting in transition from euchromatin to heterochromatin. One such loci mediating heterochromatin formation via NcRNA–polycomb protein association is located in the p14/p15/INK4 locus expressing signal ANRIL in humans. Another NcRNA, Fendrr, plays a pivotal role in cardiac differentiation and development. Fendrr binds to DNMT TrxG/MLL and regulates the expression of genes related to heart cells via PRC2 and G9a promoter regulation. This transcript recruits the DNMT Dnmt1, which has been found to localize to CpG islands in imprinted genes. Similarly, miR-214 reduces Ezh2, a histone methyltransferase, thus promoting skeletal muscle expression and differentiation.

Short NcRNAs are the key players in cell-specific gene silencing, interacting with the site via a stem loop structure at the site of transcription and thus the repressing gene in *cis*. NcRNA-mediated histone demethylation has also been observed in some specific loci; HOTAIR is involved in removing the activating mark H3K4me3 by recruiting demethylase Fbxl10.

The role of NcRNA in the tumorigenic epigenome has been researched quite extensively. Decreased expression of miR-449a enhances production of HDACs 1–3 and thus leads to decreased proliferation and tumor formation of hepatocellular carcinoma (HCC) cells. Oncogene-mediated downregulated microRNAs (miRNAs) involved in tumorigenesis utilize EZH2 as a mediator, resulting in EZH2 overexpression. In *Caenorhabditis elegans*, lymphoblastic leukemia model gene expression was controlled through endogenous short interfering RNAs (endo-siRNAs). Epigenetic hypermeythylation of hsa-mir-9-1 was found to be a prognostic biomarker for breast carcinoma, whereas repression of miR-148a and miR-34b/c induced metastatic events in cells.

### 2.9.1.4 Cross Play of NcRNA, Histone Modifications, and DNA Methylation in Cancer

The most classic example of a cumulative effect has been observed in cancer epigenetics. Hypomethylation of repeats and pericentromeric regions results in genomic instability, whereas promoter hypermethylation leads to transcriptional inactivation. miRNA-145, a well-characterized miRNA, has been downregulated mainly because of aberrant DNA methylation of its promoter. DNA methylation and histone modification–associated promoter silencing of miRNAs (hsa-miR-9, hsa-miR-129, and hsa-miR-137) was found to be a cause of reduced expression in colorectal cancerous tissues, with only rare occurrence in normal tissue.

Simultaneously, hypoacetylation of histones tends to up- or downregulate gene expression in tumors. The transcriptional inactivation due to promoter hypermethylation tends to affect major cellular pathways such as DNA repair, cell cycle control, Ras signaling, apoptosis, metastasis, detoxification, hormone response, and vitamin response, to name a few (Kanwal and Gupta 2012).

### 2.9.1.5 Epigenetic Silencing

Human chromosomes contain two copies of every gene—one inherited from either parent. Usually only one copy of the gene is switched on and the other remains switched off by a process called imprinting. The epigenome tends to distinguish between the imprinted gene and the switched-on copy of the chromosome.

Females tend to have two X chromosomes, and dosage compensation is an essential phenomenon mediated by X chromosome inactivation (XCI). The interplay of XIST (a conserved 17 kb long noncoding rNA (lncRNA)) and its antisense partner, TSIX (also noncoding), at the regulatory element X inactivation center (Xic) is important for the inactivation of the X chromosome. A short repeat RNA (RepA) inherent to XIST recruits PRC2, resulting in H3K27 marker trimethylation and progressive silencing of the X chromosome. Methylated H3K27 tends to accumulate PRC1 and subsequently leads to the X inactivation center.

Abnormal imprinting has been found to be associated with Beckwith–Wiedemann syndrome (characterized by body overgrowth and an increased risk of cancer), Prader–Willi syndrome (characterized by poor muscle mass with constant hunger) and Angelman syndrome (characterized by intellectual disability and motion difficulties). The imprinted human chromosome 11p15.5 (involved in Beckwith–Wiedemann syndrome) is controlled with paternally expressed antisense NcRNA; this 91-kb-long transcript facilitates bidirectional silencing by recruiting HMTs of

PRC2 complex and enzyme G9a. The interaction mediates an increase in repressive histone modification in the Kcnq1 domain KvDMR1/LIT1/KCNQ1OT1.

Analysis of the epigenetic modifications is equally necessary, as they convey significant information. SGS platforms are not competent in decrypting sequences with modified nucleotide bases. For example, 5-methylcytosine and 5-hydroxymethylcytosine are read as cytosine, leading to loss of epigenetic information during sequencing. Some specialized sequencing methods such as bisulfate sequencing and oxidative bisulfite sequencing have been developed, which enable us to distinguish regions in the genome that are prone to epigenetic changes.

Bisulfite reacts differentially with cytosine and 5-methylcytosine; cytosine becomes deaminated in the presence of bisulfite to uracil and 5-methylC is unreactive (i.e., read as C). Sequencing performed on methylated strands and unmethylated strands present distinct results. This technique holds true for dsDNA, as after treatment with bisulfite, the strands become noncomplementary and hence are treated as single-stranded DNA (ssDNA). The drawback of this technique is that it cannot distinguish between 5-OH methylcytosine, another important epigenetic modification, and unmethylated cytosine. Hence, oxidative bisulfate has come up as a solution. This technique involves chemical oxidation converting 5-hydroxymethylcytosine to 5-formylcytosine using potassium perruthenate before bisulfite addition. Subsequently, bisulfate treatment subjects the formylated derivative to deformylation and deamination to form uracil. Hence, with these methods, scientists can distinguish between cytosine, 5-methylcytosine, and 5-hydroxymethylcytosine (Villota-Salazar et al. 2016; Peschansky and Wahlestedt 2014; Baylin and Ohm 2006).

## 2.10 Genomic Methods for Studying Complex Diseases

Genome-wide association studies (GWASs) are an experimental design facilitating the study of population genetics and their complex attributes. The data generated have been used to study genetic variants in samples from populations. A genetic linkage map is constructed by observing the probability of two markers being inherited together. Markers must be polymorphic; alternative forms (copy number variants) should be present among individuals so they may be distinguished among different individuals with a frequency of once every 300–500 bp. GWASs rely on linkage disequilibrium (LD), defined as the correlation of DNA variants with respect to a finite population size, mutation, recombination rate, and natural selection. LD among variants is quantified as a squared correlation ($r^2$), as it is proportional to the sample size considered for an association study between an observed genotyped and an unobserved causal variant.

One of the first techniques enabling GWASs is a SNP array. Two platforms have been frequently used for GWASs: Illumina (San Diego, CA, USA) and Affymetrix (Santa Clara, CA, USA). Affymetrix is a chip-based platform with printed DNA probes as spots enabled to detect a distinct SNP allele. Illumina, on the other hand, is bead based and immobilized with longer DNA probes with higher specificity, but it is expensive. At the same time, detection of millions of SNPs becomes a hurdle with these techniques. This was made feasible with the use of chip-based microarray technology. In the next section we discuss the details of these techniques (Visscher et al. 2017).

### 2.10.1 ChIP and ChIP-on-Chip

Our understanding of the biological basis of interactions of DNA with nuclear proteins in the background of gene expression, cell differentiation, or disease has been hugely improved by the arrival of chromatin immunoprecipitation (ChIP). ChIP selectively analyzes a protein on the basis of its immunodomains from a chromatin preparation and thus determines the DNA sequences linked to the protein. Briefly, ChIP is an amalgamation of cross-linking and cell lysis followed by nucleic acid shearing and antibody-based immunoprecipitation (IP), DNA sample clean-up, and amplification by PCR in tandem. Cells fixed with formaldehyde are subjected to cell lysis. Sonication tends to result in variable batches of fragments, whereas use of micrococcal nuclease (MNase) has tended to give consistent results. Chromatin fragments may vary from 200 to 1000 bp, depending on the type of technique used for shearing. Quantification is generally done by qPCR, which is an accurate, gel-free system for the measurement of DNA enrichment. ChIP-on-chip is a variant whereby the IP technique is followed by a DNA microarray. DNA sequences linked with the precipitated protein can be recognized by end-point PCR, qPCR, labeling, and hybridization to genome-wide or tiling DNA microarrays (ChIP on-chip), molecular cloning, and sequencing (Yan et al. 2016; Ho et al. 2011).

## 2.11 Clinical Genomics

The techniques discussed earlier generate huge data which, if not evaluated properly, may lead to false negatives. One of the biggest challenges in these studies is accurate and reproducible analysis of the resulting terabytes of data. Computational analysis becomes mandatory and requires aligning of reads to reference genomic sequences. Commonly used programs include BWA for DNA reads, STAR for RNA-seq data, and Bismark, BSMAP, or BSmapper for bisulfite sequencing data. Many aspects of patient care integrate genomics and informatics, especially with the transition to electronic health records (EHRs). Large-scale studies using machine learning and data-mining methods permit unprecedented access to large sample sizes and diverse patient cohorts, facilitating studies related to adverse drug effects and developing a classifier for disease phenotype severity.

Genomic approaches have been instrumental in management of chronic illnesses, such as diabetes and inflammatory bowel disease. The Human Microbiome Project and other metagenomic studies have disclosed the composition of microbiota and their adverse effects. Fecal microbiota transplantation for treating *Clostridium difficile* is one example representing the translation of this finding into clinical practice.

Methylated DNA sequences are one of the significant biomarkers for prognosis and diagnosis of various anomalies. Epigenetic markers have been evaluated in tumors and body fluids. For example, hypermethylated *CDH13*, *MYOD1*, *MGMT*, *p16 $^{INK4a}$*, and *RASSF1A* gene frequency fluctuates prominently within cancer types. Additionally, these changes are detectable in plasma DNA and urine. In the near future, diagnostic hypermethylation assays for *RASSF1A*, *RARβ2*, *APC*, and *GSTP1* may be successfully utilized for differentiation between benign and metastatic changes of the prostate.

Epigenetic drugs include DNMT inhibitors and HDAC inhibitors. Epigenetic therapies could be made possible on the basis of the "methylogenomics" now available for various malignancies. Correlation studies of the clinical outcomes when they are subjected to epigenetic drugs have already been studied in detail. A classic success story regarding the use of genomics in cancer therapy relates to use of the BRAF inhibitor vemurafenib in metastatic melanoma. Genomic screening confirmed BRAF V600 mutations in 50% of patients, which increased the sensitivity of cancer cells to BRAF inhibitors. These models will be essential for early intervention in individuals at high risk of different malignancies. Currently, the legal and ethical issues surrounding clinical genomics, including genetic testing in children and adolescents, are been evaluated.

Federal policy changes this evolution in our understanding and treatment of cancer, most notably through US President Obama's Precision Medicine Initiative, announced in his 2015 State of the Union Address. This initiative includes increased funding for the National Cancer Institute to research genomic drivers in cancer and to streamline the design and testing of targeted therapies based on genetics. Relatedly, the prototypical clinical trial is transforming to reflect a personalized medicine approach, as seen by the success of the IMPACT and IMPACT2 studies. Importantly, these changes in clinical genomics are occurring on a global scale, inspiring international cooperation to advance medicine (Vijay et al. 2016).

## 2.12   Genomic Databases

Biological databases offer researchers access to extensively mined biologically relevant data, encompassing the genomic sequences of an increasing variety of organisms. The data from these repositories may be reviewed further for a plethora of clinical applications. The Table 2.3 provides a comprehensive compilation of databases available for genomic studies.

**Table 2.3** Genomic databases

| Database | URL | Importance |
|---|---|---|
| Genome Database (GDP) | http://www.gdb.org | Compilation of human genes, clones, STSs, polymorphisms, and maps |
| Database of Genomic Variants (DGV) | http://dgv.tcag.ca/dgv/app/home | Curated data on genomic structural variations |
| Cooperative Human Linkage Center (CHLC) | http://www.chlc.org | Comprehensive human linkage map with centimorgan density |
| Unigene | https://www.ncbi.nlm.nih.gov/unigene/ | Differential expression data based on tissue, age, and health status |
| Database of Sequence Tagged Sites (dbSTS) | https://www.ncbi.nlm.nih.gov/dbSTS/index.html | Contains information about sequence tagged sites |
| Mouse Genome Database (MGD) | http://www.informatics.jax.org | Mouse genomic repository with an interface |
| Online Mendelian Inheritance in Man (OMIM) | http://www3.ncbi.nlm.nih.gov/Omim/ | Compendium of human genes and genetic phenotypes |
| Ensembl: Human Genome Central | http://www.ensembl.org/genome/central/ | Genome information at the sequence level |
| Tomato Functional Genomics Database (TFGD) | ted.bti.cornell.edu/ | RNA-seq data sets with expression information |
| ArrayExpress | https://www.ebi.ac.uk/arrayexpress/ | Functional genomics data from microarray and sequencing platforms |
| Gene Expression Omnibus (GEO) | https://www.ncbi.nlm.nih.gov/geo/ | Storage and distribution of microarray and next-generation sequencing |
| ARCHS4 | https://amp.pharm.mssm.edu/archs4/ | Sequencing database for human and mouse experiments from GEO and SRA |
| Sequence Read Archive (SRA) | https://www.ncbi.nlm.nih.gov/sra | Sequencing database from high-end sequencing platforms |
| Nucleic Acid–Protein Interaction Database (NPIDB) | http://npidb.belozersky.msu.ru/ | Database for nucleoprotein complexes |
| DNAproDB | http://dnaprodb.usc.edu/ | Database of DNA–protein complexes |
| EDGEdb | http://edgedb.umassmed.edu | *Caenorhabditis elegans* transcription factor–DNA interaction data based on differential gene expression |
| Protein–DNA Interface Database (PDIdb) | http://melolab.org/pdidb/web/content/links | Structural information on protein–DNA interface based on x-ray crystallography |
| NASCArrays | http://ssbdjc2.nottingham.ac.uk/narrays/experimentbrowse.pl | Single- and double-channel microarray experiments for *Arabidopsis* |
| ONCOMINE | https://www.oncomine.org/ | Cancer microarray repository data-mining platform |
| GENEVESTIGATOR | https://www.genevestigator.ethz.ch. | Gene expression profiles of more than 22,000 *Arabidopsis* genes |

(continued)

**Table 2.3** (continued)

| Database | URL | Importance |
|---|---|---|
| Soybean Genomics and Microarray Database (SGMD) | http://psi081.ba.ars.usda.gov/SGMD/default.htm | Genomic, EST, and microarray data on soybean–nematode interaction and embedded analytical tools |
| Yale Microarray Database (YMD) | https://medicine.yale.edu/keck/ymd/ | Database for archiving and retrieving microarray data generated by Affymetrix, Illumina, Nimblegen, and Sequenom |
| PLAZA | https://bioinformatics.psb.ugent.be/plaza/ | Integrative database for plant functional, evolutionary, and comparative genomics |
| HOBACGEN | http://pbil.univ-lyon1.fr/databases/hobacgen.html | Database for comparative genomics in bacteria |
| MethylomeDB | www.neuroepigenomics.org/methylomedb/ | Brain methylome database supplementing DNA methylation profiles from humans and mice |
| DiseaseMeth | bio-bigdata.hrbmu.edu.cn/diseasemeth/ | Methylomes of human disease |
| HIstome | www.actrec.gov.in/histome/ | Histones with their modification sites, variants, and enzymes that mediate alteration |
| TarBase 6.0 | diana.imis.athena-innovation.gr/DianaTools/ | Data on miRNA targets |
| NONCODE v3.0 | www.noncode.org | Microarray-based expressional and functional lncRNA data |
| miRNEST | mirnest.amu.edu.pl/ | Repository of animal, plant, and virus miRNA data |
| Clinical Genomic Database (CGD) | https://research.nhgri.nih.gov/CGD/ | Database of disorders with known genetic predispositions and available interventions |
| ClinGen | https://www.clinicalgenome.org/ | Database of clinically relevant variants of genes important for precision medicine and research |
| Canadian Open Genetics Repository (COGR) | http://opengenetics.ca/ | Classification of human genetic variants of all kinds and resources |
| DBTBS | http://dbtbs.hgc.jp | Transcriptional regulation events in *Bacillus subtilis* |
| WormBase | http://www.wormbase.org/ | Data repository for information about *Caenorhabditis elegans* and related nematodes |
| GreenPhylDB | http://greenphyl.cirad.fr | Comparative functional genomics in rice and *Arabidopsis* genomes |
| MolliGen | http://cbi.labri.fr/outils/molligen/ | Comparative genomics platform for Mollicutes |
| The Adaptive Evolution Database (TAED) | http://www.bioinfo.no/tools/TAED | Phylogeny-based tool for comparative genomics |

*EST* expressed sequence tag, *lncRNA* long noncoding RNA, *miRNA* microRNA, *seq* sequencing, *STS* sequence-tagged site

# References

Baylin, S. B., & Ohm, J. E. (2006). Epigenetic gene silencing in cancer—a mechanism for early oncogenic pathway addiction? *Nature Reviews Cancer, 6*(2), 107–116.

Cavalli-Sforza, L. L. (2005). The human genome diversity project: Past, present and future. *Nature Reviews Genetics, 6*(4), 333.

Chial, H. (2008). DNA sequencing technologies key to the Human Genome Project. *Nature Education, 1*(1), 219.

ENCODE Project Consortium. (2012). An integrated encyclopedia of DNA elements in the human genome. *Nature, 489*(7414), 57.

Haubold, B., & Wiehe, T. (2004). Comparative genomics: Methods and applications. *Naturwissenschaften, 91*(9), 405–421.

Ho, J. W., Bishop, E., Karchenko, P. V., Nègre, N., White, K. P., & Park, P. J. (2011). ChIP-chip versus ChIP-seq: Lessons for experimental design and data analysis. *BMC Genomics, 12*(1), 134.

Jia, M., Guan, J., Zhai, Z., Geng, S., Zhang, X., Mao, L., & Li, A. (2017). Wheat functional genomics in the era of next generation sequencing: An update. *The Crop Journal, 6*(1), 7–14.

Kanwal, R., & Gupta, S. (2012). Epigenetic modifications in cancer. *Clinical Genetics, 81*(4), 303–311.

Kchouk, M., Gibrat, J. F., & Elloumi, M. (2017). Generations of sequencing technologies: From first to next generation. *Biology and Medicine, 9*(3), 3–8.

Koonin, E. V., & Galperin, M. Y. (2003). Genome annotation and analysis. In *Sequence—evolution—function* (pp. 193–226). Boston: Springer.

Peschansky, V. J., & Wahlestedt, C. (2014). Non-coding RNAs as direct and indirect modulators of epigenetic regulation. *Epigenetics, 9*(1), 3–12.

Snyder, M. W., Adey, A., Kitzman, J. O., & Shendure, J. (2015). Haplotype-resolved genome sequencing: Experimental methods and applications. *Nature Reviews Genetics, 16*(6), 344.

Tessarz, P., & Kouzarides, T. (2014). Histone core modifications regulating nucleosome structure and dynamics. *Nature Reviews Molecular Cell Biology, 15*(11), 703.

Vijay, P., McIntyre, A. B., Mason, C. E., Greenfield, J. P., & Li, S. (2016). Clinical genomics: Challenges and opportunities. *Critical Reviews™ in Eukaryotic Gene Expression, 26*(2), 97.

Villota-Salazar, N. A., Mendoza-Mendoza, A., & González-Prieto, J. M. (2016). Epigenetics: From the past to the present. *Frontiers in Life Science, 9*(4), 347–370.

Visscher, P. M., Wray, N. R., Zhang, Q., Sklar, P., McCarthy, M. I., Brown, M. A., & Yang, J. (2017). 10 years of GWAS discovery: Biology, function, and translation. *The American Journal of Human Genetics, 101*(1), 5–22.

Waterson, R. H., Lander, E. S., & Wilson, R. K. (2005). Initial sequence of the chimpanzee genome and comparison with the human genome. *Nature, 437*(7055), 69.

Wei, L., Liu, Y., Dubchak, I., Shon, J., & Park, J. (2002). Comparative genomics approaches to study organism similarities and differences. *Journal of Biomedical Informatics, 35*(2), 142–150.

Yan, H., Tian, S., Slager, S. L., Sun, Z., & Ordog, T. (2015). Genome-wide epigenetic studies in human disease: A primer on -omic technologies. *American Journal of Epidemiology, 183*(2), 96–109.

Yan, H., Tian, S., Slager, S. L., & Sun, Z. (2016). ChIP-seq in studying epigenetic mechanisms of disease and promoting precision medicine: Progresses and future directions. *Epigenomics, 8*(9), 1239–1258.

# Chapter 3
# Transcriptomics

**Jyotika Rajawat**

**Abstract** Transcriptomics can be considered as Integromics, whereby combining data from various omic branches results in crisp information. Transcriptomics provides the most informative base to start a research work, and with advent of new high-throughput techniques, it has become very easy and fast to generate a pool of data and information. Transcriptome constitutes all transcripts present in a cell including mRNA, miRNA, noncoding RNAs, and small RNAs. Transcriptomics identifies the quantity of RNA and transcriptional structure and quantifies the differential expression levels of transcripts spatially and temporally during various developmental stages and under varying physiological conditions. It gives the information on diversity, noncoding RNAs, and the arrangement of transcriptional units in coding regions. Transcriptomic analysis began with a primitive technique called EST, i.e., expressed sequence tags, followed by another technique called SAGE, i.e., serial analysis of gene expression, based on Sanger sequencing. EST and SAGE were laborious and determined a small set of transcripts in a random fashion, yielding half information on transcriptome. The 1990s marks the revolutionary decade in transcriptomics with the introduction of technological innovation of contemporary technique called microarray. Microarray analyzes large mammalian transcriptome rapidly and has been useful in drug development and clinical research by analyzing thousands of genes from multiple samples. The major drawback of the technique is the analysis of only known sequences and hence cannot detect novel transcripts. The latest in transcriptome analysis is RNA-Seq based on deep sequencing technology which can record up to $10^9$ transcripts. It identifies the gene and the temporal activity of genes in a genome. In situ RNA-Seq is an advanced form which gives an overview of an individual cell in a fixed tissue. RNA-Seq is thus an advanced technique providing detailed information of complex transcriptome. Further, the chapter discusses the advantages and limitations of transcriptome analysis tools.

J. Rajawat (✉)
Molecular and Human genetics laboratory, Department of Zoology, University of Lucknow, Lucknow, Uttar Pradesh, India
e-mail: jrajawat@gmail.com

© Springer Nature Singapore Pte Ltd. 2018
P. Arivaradarajan, G. Misra (eds.), *Omics Approaches, Technologies And Applications*, https://doi.org/10.1007/978-981-13-2925-8_3

Similarly, the information of the expressed genes from a microbial community is termed as metatranscriptomics. Metatranscriptomics provides functional profile of microbiome under varying physiological conditions. The data generated is useful in enrichment analysis and phylogenetic analysis of microbes. Several bioinformatic pipelines are now designed or are in process for analysis of metatranscriptome dataset. Metatranscriptomics will provide information on microbial flora in human beings which can be exploited for designing targeted therapy for microbial dysbiosis. The last section of the chapter discusses the application of transcriptomics particularly in diagnosis and profiling a disease. Another application includes identifying environment-responsive genes or pathways, host-pathogen interactions, and annotating gene functions.

**Keywords** EST · SAGE · Microarray · RNA-Seq · Gene annotation · Cancer

# 3.1 RNA to Transcriptome

## *3.1.1 Transcriptome and Transcriptomics*

A classic simple transcriptome is composed of spliced mRNAs each consisting of 5′ capped end, a 5′ UTR, and a coding sequence (CDS) followed by 3′UTR ending in polyA tail. The above definition offers a simple form of transcriptome, but there has been a fundamental complexity linked to gene to protein transformation. Complexity can be observed at multiple levels, starting from alternative splice variants coded from a single gene leading to isoforms of protein with functional diversity and redundancy. Another complexity often observed is noncoding of transcripts to protein as they lack CDS. Diversity is also seen in 5′ and 3′ ends; thereby, different regulatory mechanisms are followed by differential mRNA turnover. mRNA levels vary depending upon the degree of transcription and their stability. Transcriptome connects genome to gene function. Analyzing the transcriptome of an organism would yield an overview of an expressed gene and will be highly informative for understanding development and disease using other approaches like proteomics and metabolomics.

Analysis of a complete set of transcripts in a cell or transcriptome structure at a given time is known as transcriptomics. The study of the RNA, RNA variants, transcriptome complexity, and gene expression analysis is termed as transcriptomics. Transcriptomics guides us to analyze all RNA transcripts including noncoding RNAs (ncRNAs) and small RNAs, splicing variants and pattern, transcriptional units, transcriptional start sites, and posttranscriptional modifications. We can analyze the differential expression of gene population under different conditions and developmental stages. With the advancement of techniques, differential gene expression in a spatial and temporal manner can also be now analyzed with transcriptomics. Transcriptomics is now the first and foremost assay to understand an organism's biology where it reveals the information on expression of a gene, its regulation, and downstream signaling.

## 3.1.2  Principle of Transcriptomics

Structure and dynamics of the transcriptome is analyzed by transcriptomics approach. Transcriptomics is based on the analysis of mRNA, where mRNA is converted to cDNA followed by fragmentation, labeling, hybridization, and probing. Sequencing is then carried out based on NGS followed by bioinformatic approach for data analysis.

## 3.1.3  Technological Approach to Study Transcriptomes

The word transcriptome was first coined in the early 1990s. The earliest technique developed to study transcriptome was Sanger sequencing-based method known as EST and SAGE. The conventional techniques were later replaced by contemporary techniques microarray and RNA-Seq. In this section we describe all techniques used to study transcriptomes. The initial material for all these methods is RNA enriched with mRNA. RNA is extracted from cells and tissues using TRIzol Reagent and finally eluted using polyA affinity columns for mRNA enrichment. Care should be taken to avoid the contamination of RNase enzyme during RNA isolation.

### 3.1.3.1  Expression Sequence Tag (EST)

EST is a technique which generates short oligonucleotide sequence based on the principle of Sanger sequencing method. RNA is transcribed to cDNA using reverse transcriptase enzyme, and then cDNA generated is sequenced. ESTs can be generated from any mixture of samples of any organism as it does not require prior knowledge of the origin of sample. Just as sequence-tagged sites (STS) mark the genomic DNA, similarly ESTs are unique sequences pointing to expressed genes in the mapped cDNA clone. EST was useful for unknown gene identification but could not quantify expressed genes. Another major drawback of EST was sequencing a single cDNA copy at a time, making it low throughput and costly method. With the advent of new technique, EST is no longer used, but EST libraries have been the basis for designing of early microarray gene chips.

### 3.1.3.2  Serial/Cap Analysis of Gene Expression (SAGE/CAGE)

An advancement of EST was serial analysis of gene expression (SAGE) where fragments were tagged which allowed quantitation of transcripts. SAGE was invented in 1995 by Dr. Victor Velculescu from John Hopkins. SAGE is a powerful technique designed for direct quantitation (digital analysis) of gene expression and also identifies novel gene expression in a cell population. Basic principle underlying SAGE is:

(a) Simultaneous analysis of thousands of genes by tagging short oligo sequences of 9–10 bp at 3'end for each transcript.
(b) Sequence tags are linked together, then cloned, and sequenced. Serial and parallel analysis can be performed together thereby increasing the data output.
(c) Quantitation is finally done, wherein each tag count represents the abundance of a particular transcript.

**Method**

1. cDNA synthesis-mRNA is converted to cDNA using biotinylated oligo dT primer which is then digested with restriction enzymes known as anchoring enzyme (AE) which cleaves at every 256 bp generating sticky ends. Commonly used anchoring enzyme is NlaIII which is 4 bp recognizing enzyme. Biotinylated 3'cDNA binds to streptavidin-coated beads and is affinity purified from the pool of RNA. These captured cDNA mixture is then divided into two halves, and each half is ligated with specific independent linkers (A or B) through NlaIII cohesive ends. Linkers are docking molecules made up of oligonucleotide duplex. Linkers contain NlaIII overhangs, recognition sequence for tagging enzyme (TE), and primer sequence A/B. The mixture is now digested with type IIS tagging enzyme which is normally BsmFI or FokI, releasing linker-adapted SAGE tags with staggered ends. Staggered ends are converted to blunt ends with the help of Klenow DNA polymerase. The resultant products are short tags from each transcript. The two divided pools are now mixed followed by ligation in tail-tail orientation with T4 DNA ligase forming ditags sandwiched between linkers. These are PCR amplified followed by digestion with anchoring enzyme NlaIII resulting in separation of linkers and leaving sticky end ditags which are finally separated on gel (PAGE). All such ditags separated are isolated from the gel and ligated to form long molecules termed as concatemers. Concatemers are cloned into vector, amplified in bacteria, and a large number of copies are isolated and then sequenced (Velculescu et al. 1995; Moreno et al. 2001; Yamamato et al. 2001). Sequencing results into a vast data in a form of long nucleotide chain which is analyzed with the SAGE software.

The presence of AE site between two tags allows the software to identify the end of one tag and beginning of the other tag. The software further analyzes the number of tags, determines the abundance of the same tag from the single transcript, and identifies whether the tag belongs to a known gene or is novel. SAGE software aligns the data with the GenBank sequence and identifies it (Fig. 3.1).

**Limitations**

(a) Actual gene expression cannot be measured.
(b) Quantitative bias is observed due to linker dimer molecular contamination, low-efficient ligation of blunt ends, or amplification artifact.
(c) Sequencing errors could lead to difficulty in assigning tag to a specific transcript.
(d) There could be same tag for two different genes or different tags for splice variants of the same gene.

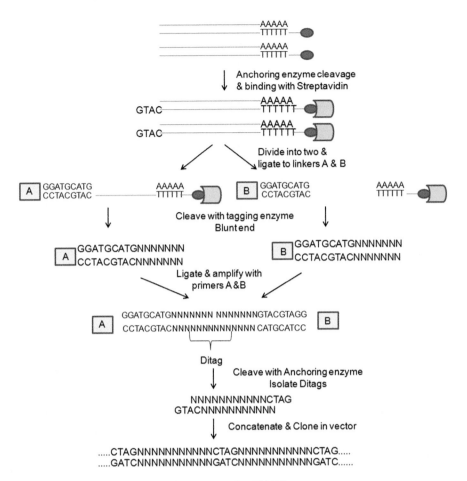

**Fig. 3.1** Steps for serial analysis of gene expression (SAGE)

## Applications

1. Yeast: The whole transcriptome profile of yeast has been done with SAGE profile where 60,633 tags correspond to 4665 genes and out of which 93% matches with the yeast genome and the expression level varies from 0.3 to 200/cell.
2. Cancer: SAGE analysis has been highly exploited for cancer studies where differential gene expression profiles of normal and cancer cells are determined.
3. Tissue analysis: Transcriptome of various tissues including renal, cervical, etc. has been analyzed with SAGE.
4. Immunological studies: Differentially expressed genes in response to immunoglobulin E high-affinity receptors have been identified. Several genes have been identified as stimulation-responsive genes with the help of SAGE analysis. Differential gene expression in response to different colony-stimulating factors was also analyzed with SAGE profile. Differential gene expression response in monocytes and dendritic cells was characterized by SAGE technique (Yamamato et al. 2001).

### 3.1.3.3 Microarray

Microarray technique has brought revolution in the field of molecular biology which opened up the avenue to study biological functions of related genes to global gene expression and pathway analysis. It yields a massive amount of data which can be deciphered to a sea of information about biological activity of a cell. Microarray was developed to monitor multiple gene expressions in a given time simultaneously. Scientific ingenuity has led to transition from two-dimensional to three-dimensional microarray followed by suspension bead arrays, which are now useful in clinical implications. The technique originated basically from large-scale mapping of genomic DNA and sequencing, later adapted for global transcript analysis crediting the huge success of microarray. Several other types of microarray are in progressive state or are under development that would change the future of research and medical treatment (Hoheisel 2006).

The basic principle of microarray is hybridization between complementary strands of DNA where one DNA strand (short oligos) or probes are arrayed on microchip and the fluorescent labeled target transcript is added. Fluorescent intensity of each probe determines the transcript abundance of a particular gene or RNA, and the position on the chip identifies the target. Thus, microarray can produce quantitative data yielding information about gene expression or qualitative data useful for diagnostic purpose. Microarray chip consists of immobilized phase, i.e., probes which could be cDNA prepared from EST library or genome sequence or oligonucleotides. DNA is then snap dried and UV cross-linked to the glass surface of an array. The mobile phase in RNA microarray is usually labeled cDNA. Microarray of transcriptome is basically cDNA/oligo array which is classified into two types:

(a) Low-density spotted arrays or printing microarrays – picoliter volume of cDNA is required, and control and treated samples are labeled with different fluorophores and hence can be incorporated in same array. The oligonucleotide probes (50–70 nucleotide in length) or PCR products are spotted onto the glass slides with spot size ranging from 80 to 150 μm. The information obtained from these arrays is the relative gene expression between two conditions as absolute quantification cannot be done.

(b) In situ (on-chip) synthesized or high-density arrays or gene chips consist of synthetically designed oligonucleotides on glass slides or wafers with the help of modified photolithographic technology (by Affymetrix) or inkjet technology (Agilent). The probes are usually 20–25 bp long, and multiple probe sets are used that enhance the sensitivity and specificity of the array compared to low-density array. Hence, a single gene is assayed by several short oligo probes. The major advantage of these gene chips is absolute measurement of the RNA expression in each sample, and the main drawback is inability to simultaneously compare two biological samples in a same array (Lowe et al. 2017).

## Method

Total RNA is isolated from biological samples, and purity is checked and quantified. Required amount of RNA is then processed for tailing using ATP mix, and polyA RNA is generated. Tailed RNA is then ligated with the FlashTag Biotin HSR label using T4 ligase. Hybridization cocktail is prepared using tagged ligation mix and hybridization mix reagents and then injected in the microarray chip followed by hybridization in the oven overnight. Hybridization is carried out at 48 °C and 60 rpm for 16 h. During hybridization the target transcripts bind to the respective probes. After completion of hybridization, arrays are removed from the oven and filled completely with array holding buffer for equilibration. Further, arrays are washed and stained with respective buffers and staining solutions for 2–3 cycles as per the fluidic station protocol and array format being used. Finally, array is filled with Array Holding Buffer and ensures no air bubbles are trapped in the array. Seal the array with septa, wipe it properly, and place it in the scanner for further analysis. Data is then summarized using a software, data quality determined from spike-in controls and normalized with internal control (Fig. 3.2).

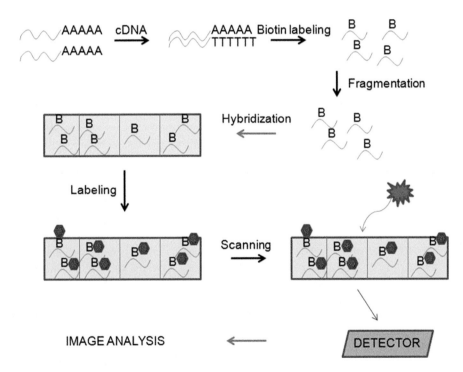

**Fig. 3.2** Flow chart depicting the method for microarray

**Limitations**

Microarray identifies already reported genes and cannot predict novel genes. Analysis of the result also remains a challenge as normalization process is affected by technological variation more than biological differences.

**Applications**

(a) Gene expression profile of a cancer cell (Govindarajan et al. 2012).
(b) Microarray has identified molecular signatures specific for cancer types and subtypes. Based on the microarray gene signature analysis, acute myeloid leukemia was distinguishable from acute lymphocytic leukemia (Golub et al. 1999). Lately, another group has characterized molecular signature in nonneoplastic and neoplastic prostate tissues to differentiate healthy prostate, neoplastic prostate, localized cancer, and metastatic prostate cancer (Dhanasekaran et al. 2001). Now, signature profiles of many cancers have been reported by various groups. Tumor-specific molecular markers have also been reported with the use of microarray technique.
(c) Differentiate normal, precancerous, and cancerous cells.
(d) Inflammatory signature.
(e) Drug response profile: Drug resistance is a common problem being faced with multiple cancer drugs, and hence using microarray technology to identify the cellular pathways implicated in resistance could help in overcoming the drug resistance phenomenon (MacGregor and Squire 2002).
(f) Comparative analysis to identify gene sets responsible for high antibody production in mammalian cells (Yee et al. 2008).

### 3.1.3.4   RNA-Seq

RNA sequencing is a novel method applied for mapping and quantifying the transcriptome. RNA sequencing is an advanced technique which utilizes deep sequencing approach to analyze the transcriptome. The major advantage of RNA-Seq over microarray is the discovery of novel RNA species. RNA quantification is done at single-base resolution and is a cost-effective high-throughput analysis of transcriptome. RNA-Seq offers higher sensitivity and dynamic range where a broad range of expression is captured; microarray exhibits saturation, while RNA-Seq at extreme values is in linear scale. RNA splice events can be detected by RNA-Seq while not with microarray (Wang et al. 2009; Wolf 2013).

**Method**

RNA sequencing is possible, but most of the instruments are based on DNA sequencing, and hence cDNA library preparation is a required and critical step. As mentioned above that before reverse transcription, polyA transcript is selected, unwanted RNA (ribosomal RNAs) is depleted, and further selected RNA samples are fragmented to smaller size due to size limitations in sequencing platforms. Alternatively, cDNA can also be fragmented by acoustic shearing (sonication) or using DNase I. Recently, another approach known as tagmentation has also been developed based on transposon where Tn5 transposase fragments the cDNA and simultaneously ligates the adapter oligos at both ends. Fragmented full-length cDNA or cDNA generated from fragmented RNA is ligated with adapters. Ligation of adapter results in lack of strand specificity, hence making the directionality of RNA strands difficult to predict. Several approaches have been developed to provide directionality; using different adapters at both ends is one such approach. Another method utilizes incorporation of dUTP in second-strand cDNA which can be degraded using uracil-DNA glycosylase (UDG) before the amplification step. Hence, only the first strand with defined adaptor is amplified. Before sequencing, the cDNA libraries have to be amplified by PCR using 8–12 cycles. Uneven amplification results due to difference in size and composition of cDNA, and the issue is addressed by using unique molecular identifiers (UMIs) which distinguish PCR products from the artifacts. These molecular labels can be introduced in RNA during RT or in adapter sequences or by Tn5 transposase during cDNA fragmentation. Molecular labeling with UMIs is of great significance in single-cell RNA-Seq where input RNA is in very low quantity (Hrdlickova et al. 2017; Lowe et al. 2017).

Information about transcript is not retrieved as a whole but is generated as short reads of several hundred base pairs. If the transcript information is available, then the read sequence is aligned to the reference; but if the transcriptome information is not available, then de novo assembly is carried out for the reads or read pairs. Advancement in bioinformatic data analysis has made it possible to obtain novel sequence information from the sequencing data of several individuals (Fig. 3.3).

**Fig. 3.3** RNA-Seq. mRNA is isolated from a cell/tissue, fragmented, converted to cDNA, library constructed, sequenced, and mapped with reference or de novo assembled

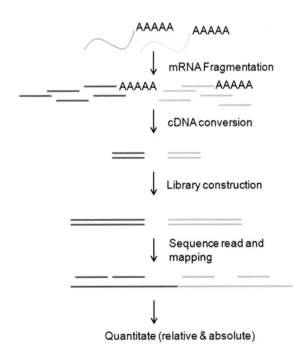

**Challenges for RNA-Seq**

Sequencing biases or library construction:

The manipulations done during cDNA library preparation can limit its usage in whole transcript profiling. Different fragmentation method creates a bias product which is a problem in identifying strand specificity and transcript orientation. Strand-specific library construction is another issue of concern as it is a laborious process to carry out.

Bioinformatics:

The short reads are assembled into contigs and then aligned to reference genome, and transcriptional structure is revealed. There are several programs that map the reads to the genome sequence like ELAND, RMAP, MAQ32, and SOAP31. PolyA tail reads can be identified with multiple A's at the end, and exon-exon junction can be identified by specific sequence flanking the splice site. Difficulty to identify arises for reads that span splice junction due to alternative or trans-splicing. Thus, there is a need to develop computational program to identify novel splicing methods (Wang et al. 2009).

Larger genome offers more complexity and hence more sequencing depth for sufficient coverage.

Different origin for comparing the sample RNA and the reference genome.

Presence of transcriptional background noise due to incompletely processed RNAs.

**Applications**

1. With RNA-Seq many novel transcribed regions and splicing isoforms of genes have been revealed.
2. 5′-3′ exon boundaries have been defined, and introns can be mapped by looking for the tags containing GT-AG splicing sites.
3. RNA-Seq is useful in examining the splicing diversity by scanning the reads for reported or novel splice junctions.
4. Several novel transcripts have been reported in *Saccharomyces cerevisiae* and *Schizosaccharomyces pombe* with the help of RNA-Seq that were undetected by microarray.
5. RNA-Seq captures transcriptome dynamics and gives digital measurement of gene expression across different tissues and varying conditions (Wolf 2013).

### 3.1.3.5 Alternative New Approach to RNA-Seq

There are several variants to RNA-Seq depending on the goal of analysis:

RNA sequential probing of target (RNA-SPOT) is an alternate, accurate, and low-cost approach to sequence transcriptome in a cell. The transcripts are captured on coverslip containing locked nucleic acids (LNA) poly-dT and hybridized with a pool of thousand transcripts, and 12 pseudocolor schemes have been used for barcoding approximately 10,000 genes for hybridization (Linus Eng et al. 2017).

Single-cell transcriptomics is used for measuring gene expression in a single cell and where single mRNA molecule is detected by using mRNA FISH probe. This will allow the count or abundance of single transcript in an individual cell (Kanter and Kalisky 2015).

Single-cell RNA-Seq (scRNA-Seq) characterizes a cell population within a tissue, and spatial transcriptomics (ST) analyzes spatial gene expression within a tissue. Moncada and colleagues have recently integrated scRNA-Seq with ST to generate a tumor atlas for pancreatic ductal adenocarcinoma where single-cell population has been mapped with spatial regions in the tissue. They identified three types of cell population in the tumor occupying specific regions (Moncada et al. 2018, BioRxiv). This approach would be beneficial in embryonic stage identification and for studying dynamic processes like bacterial infection.

## 3.2 Metatranscriptome

Microbiomes are ubiquitous and regulate the environmental health. Changes in microbial community could affect the environmental niche and the surrounding organisms as well as human beings. Thus, detail analysis of this community would help in identifying the changes and impact of environmental challenges. Metatranscriptomics (MT) describes the transcriptomic profiling of active microbes by RNA-Seq in the given environment or human niche. MT studies can capture the transcript-level changes or differential gene expression in microbial community induced by inter-/intraspecies interaction or environmental changes. MT studies would give a detailed picture of functional profile of microbiomes. There have been two major ongoing microbiome studies that involve MT studies along with metagenomic approach, human microbiome studies and environment microbiome studies. Human microbiome studies objective was to identify the correlation between changes in microbiome profiles and human diseases. This project has observed the different structure of microbes in the human mouth, airways, gut, skin, and urinogenital areas. The other part of the project is focused on gut microbiome, emphasizing on taking healthy microbiota in food and regulating health and diseases like obesity and inflammatory bowel symptoms. Environmental microbiome studies is another major project initiated in 2010 and focuses on diverse ecosystems including the ones in human, animals, plants, fresh water, marine, terrestrial, air, and all intersections of ecosystems. This project aims to characterize the structure, diversity, and distribution of microbiomes in various ecosystems on the earth (Aguiar-Pulido et al. 2016). Metatranscriptomics thus enable to monitor the change in abundance, composition, and function of gene in response to change in environment of an active fraction of microbe or microbiome.

### 3.2.1   Gene Expression Analysis

Metatranscriptomics has not been much exploited for microbiome studies as compared to metagenomics, but the study on functional analysis of microbiome is now picked up by metatranscriptomics. Shotgun RNA sequencing (RNA-Seq) allows the whole genome profiling of the active microbiome under varying conditions and hence identifies RNA-based gene regulation and biological signatures. The data obtained after RNA-Seq is analyzed by two strategies to identify the differential gene expression pattern, (a) mapping of obtained sequence reads to reference genome and (b) de novo assembly of novel transcripts.

The first strategy deals with mapping of the sequence read to the reference genome or pathway which identifies the functionality of the expressed genes and further the taxonomical classification of the microorganism. Sequence reads are aligned using NCBI database and special alignment tools like BLAST, Bowtie2, and BWA, followed by annotation using available software or resources including KEGG, GO, COG, and Swiss-Prot. The differentially up- or downregulated genes in microbiome could be identified during a disease condition. After focusing on the gene expression, finally different downstream analysis is carried out like enrichment analysis or PCA-based phylogenetic studies. Stable isotope probing (SIP) is the latest advancement in metatranscriptomics which retrieves specific targeted transcriptome of a particular microbe in an environment, like targeting aerobic microbes in lake sediment. With this strategy the relative expression of individual genes is inferred from the available database and hence is limited to the available information of reference genome in the database.

The second strategy infers the information about a gene expression from the assembled sequences collected from the short read data. The short reads of metatranscriptomic data are assembled to form longer fragments named as contigs with the help of software packages or de novo sequence assemblers. Trinity, AbySS, Trans-AbySS, MetaVelvet, SOAPdenovo, and Oases are some of the available software for de novo assembly analysis. Trinity was found to be the most efficient and sensitive, outnumbering other software programs in recovering full-length transcripts and their isoforms too (Bikel et al. 2015; Aguiar-Pulido et al. 2016). Trinity was able to reconstruct the transcripts within the highest expression quintiles. RNA-Seq by expectation maximization (RSEM) another quantitative pipeline can also be used for metatranscriptomic analysis along with Trinity for de novo assembly (Li and Dewey 2011). MEGAN is a software used for enrichment analysis by annotating the genes with GO. Simple Annotation of Metatranscriptomes by Sequence Analysis (SAMSA) is recently designed as a simple and efficient pipeline for analyzing large paired RNA-Seq datasets using supercomputing cluster (Westreich et al. 2016; 2018).

### *3.2.2  Gene Activity Diversity*

Metatranscriptomics can be used to analyze microbiomes found in diverse environment as seawater, deserts, and soil. An activity of a particular gene in diverse environment can be studied simultaneously by collecting samples from diverse regions and analyzing the functional profile. Compared to 16S rRNA gene analysis which yielded lower diversity metrics, mRNA-based analyses were with higher diversity metrics. The mRNA-derived short read data can propose significant taxonomic difference that is a reflection for difference in habitat of microbiome (Jiang et al. 2016).

Expression level polymorphism (ELP) or the differences in gene expression are responsible for phenotypic variations in numerous species. Comparing nucleotide polymorphism with ELP would help in assessing the type of genetic diversity.

## 3.3  Applications

### *3.3.1  Disease Profiling*

Transcriptomic strategies in diagnosis and profiling of several diseases have been applied in diverse field of biomedical research. Microarray and RNA-Seq approach have allowed to compare the differential gene expression in normal and disease tissue samples from patients. The strategy has been widely used in identifying cancer and in immune-related diseases.

Cancer: Cancer is caused due to genetic changes leading to altered expression of oncogenes or tumor suppressor genes. Microarray has widely been used for identifying cancer progression, classification of tumors, and drug sensitivity or resistance. Various techniques of transcriptomics are highly applicable in expression profiling of tumors and to differentiate cancerous cells from normal cells based on differential gene expression pattern. Tumor genotyping and classification are important aspects adapted from transcriptomic profile. Microarray-based expression profiling identifies cellular changes occurring during transformation of noninvasive to invasive cell undergoing metastasis. It also aids in identifying biomarkers for different cancers and have revealed certain genes responsible for chemoresistance. Transcriptome profiling is also useful in drug discovery whereby drug effects in a cell could also be monitored. Drug sensitivity and toxicity effects will be analyzed by microarrays by clinicians in the future for clinical trials. Microarray has recently been used in identification of precancerous oral lesions to be malignant or not. With the advancement of transcriptomic technology, we are entering into post-genomic era where holistic approach will be applied for personalized therapy to cancer patients. Microarray technology will be used for diagnosis, prognosis, and biomarker analysis and drug response in patients (MacGregor and Squire 2002).

Immunity and inflammation: Analysis of molecular signatures at inflammatory sites reveals the type of inflammation and help in diagnosis of infection. SAGE technique has identified differential gene expression in response to IgE or stimulating factors and has determined several novel genes to be stimulation responsive. RNA-Seq has been potentially useful in immune disorders or diseases where it can dissect two cell populations and identify T-cell and B-cell receptor repertoire in patients (Byron et al. 2016).

### 3.3.2   Ecology

Molecular ecology has progressed ahead of the use of the limited number of markers, and now whole genome or transcriptome analyses of gene expression are used to study the molecular adaptations to environmental challenges. Although transcriptomic technology has been challenging and requires technical expertise and high price, still in the last 10 years, many ecological studies have used microarrays and RNA-Seq. Transcriptomics has been applied to analyze gene expression and identification of pathways in response to abiotic and biotic environmental stresses. A set of genes responsible for biofilm formation in *Candida albicans* (fungal pathogen) were identified by RNA-Seq (Garcia-Sanchez et al. 2004). Differential gene profile in different developmental stages of chickpea in response to drought and saline stress was analyzed by transcriptomic analysis (Garg et al. 2016).

Aquatic life has been disturbed due to heavy metal contamination, oil spillage, and other pollutants in freshwater and seawater system leading to severe risk for marine life. It has become a serious global issue and needs to be handled empathically. Cadmium toxicity is one such global problem whereby it gets accumulated in aquatic animals and poses life threat to them. One such study has reported the use of transcriptome sequencing of the hepatopancreas of freshwater crab (*Sinopotamon henanese*) in response to cadmium as biomarker guide for monitoring heavy metal pollutions in water (Sun et al. 2015, Sci reports). Firstly, transcriptome dataset was generated, followed by gene annotation, and identification of genes responsive to metal toxicity, and finally cadmium altered biological pathway was characterized. The study further revealed dose-dependent cadmium effect on changes in gene expression. This is a first-in-kind study in freshwater crab and hence suggested that transcriptome datasets can be used for ecotoxicological analysis. Transcriptome profiling can be used for determining global and specific gene response to toxicity or environmental stresses and will identify markers for various species.

Transcriptome analyses have been affordable and adapted as alternative approach to lethal sampling for ecological studies. Czypionka's group has studied the thermal response on different parts of *Salamandra salamandra* by analyzing the differentially expressed genes from the tail clip and whole body. It was observed that the common thermal response was observed for approx 50% of genes and those which were not identical also belonged to similar functional types (Czypionka et al. 2015, Methods in Ecology & Evolution). Cai and group have also utilized transcriptomic technology to study the phenotypic and genotypic adaptations in Tibetan Plateau Zokors (*Myospalax*) against stressful environment. Transcriptomes from animals at three different altitudes and ecologies were sequenced and analyzed. Adaptive changes in transcriptome were observed with variation in altitudes, as under hypoxia, expression of genes like COX1 and EPAS1 were overexpressed. Genes involved in hypoxia tolerance, temperature changes, and hypercapnia tolerance were most variant (Cai et al. 2018, Sci reports). Transcriptome analysis is now being used to identify the effect of physiological stress on symbiotic relationships also. A study has observed an adaptation of a particular coral species *Galaxea fascicularis* than others in physiological stress due to chronic coastal eutrophication. The coral-algal symbiosis induces certain gene expression essential for survival and tolerance in such chronic environment (Lin et al. 2017, Sci reports). Thus, ecological transcriptomics is now an affordable and adaptable approach to study the molecular changes associated with habitat adaptation and diversification.

### *3.3.3  Evolution*

Transcriptomics in the last decade for evolutionary studies was limited only to model species whose transcript library was available for oligo array hybridization, but with development of next-generation sequencing avenues to study non-model system, evolution studies has been facilitated. De novo transcriptomics is the latest technique to study evolutionary process in non-model species whose genome information is lacking. The adaptive phenotypic variation or changes in gene expression studies would highlight how these variations occur during evolution. Gene expression variations are prominently heritable and are affected by natural selection. It also varies temporally and spatially within individual and among species. Transcriptomic technology is now being used to study the cause of phenotypic variation correlating with the divergence in the population.

Derome and colleagues used microarray technique to identify transcriptional differences in normal and dwarf fish *C. clupeaformis*. The authors reported differential expression of 51 genes mostly linked to energy metabolism, hence affecting the swimming activity of both fish types. Further analysis is carried out using classical methods like expression quantitative trail loci (eQTL) mapping or Qst-Fst tests which generate linkage map. Based on this linkage map, 34 transcripts were identified which may be under selection due to their role in divergence between two species (Derome et al. 2006, Mol Ecology).

Earlier transcriptomic analysis was focused on either the study of host or pathogen, but with RNA-Seq simultaneous transcript analysis of both host and pathogen immune interactions during infection process has been studied. Thus, dynamic response can be studied right from infection/invasion to pathogen clearance by the host immune system (Westermann et al. 2012; Durmus et al. 2015).

### 3.3.4   Gene Function Annotation

Transcriptomics has been useful in identifying the gene structure, function, and their specific phenotype. EST libraries have been used as reference for annotating genes of several closely related species. Transcriptome of *Arabidopsis* has identified metal uptake genes correlated with hyperaccumulated metals. De novo transcriptomics and advancement in RNA-Seq have enabled annotation of genes and genome of several species and threatened species like koala (Hobbs et al. 2014). RNA-Seq read assembly has been ideal for non-model organisms whose genome is poorly understood or not reported. The best example for such annotation was of Douglas fir whose database of SNPs was generated by de novo transcriptome analysis (Howe et al. 2013). Another example of gene function annotation is of genes involved in development of the muscle, cardiac, and nervous tissue in lobsters (McGrath et al. 2016; Lowe et al. 2017).

Nowadays, various online tools and databases are available for annotation process. Trinity, Blast2GO, QuickGO, AmiGO, NaviGO, REVIGO, and Gorilla are some such databases, but they have their own limitations in analysis. Gene ontology database is commonly used for functional annotation of any gene, and further functional enrichment is carried out using software or databases like UniProt, NCBI, KEGG, SEED, etc. (Lowe et al. 2017). GO FEAT is a recently developed user-friendly platform for gene annotation which is also useful for enrichment of transcriptomic and genomic dataset (Araujo et al. 2018).

**Steps for Annotation**

1. Sequencing reads were converted to FASTA format.
2. Duplicate extraneous sequence has to be removed from the FASTA file.
3. Remove sequences with poor quality N's, and also remove short reads <75 bp in length as they contain limited information; annotating these reads is difficult.
4. Remove sequences with long repeats or with 60% of single base.
5. Program is then run to generate or assemble short reads into contigs followed by transcript reconstruction.
6. These contigs with more than 500 nucleotides are then aligned or BLAST searched in NCBI database with nonredundant proteins.
7. These representative transcripts aligned with the protein database and then undergo open reading frame (ORF) prediction using TransDecoder or ORF Finder. Only those transcripts were considered full length whose ORF began and ended within the contig.
8. Finally, gene ontology was performed to identify the functional significance of the selected transcript using Blast2GO (Salem et al. 2015; Gilbert and Hughes 2011).

# References

Aguiar-Pulido, V., Huang, W., Suarez-Ulloa, V., Cickovski, T., Mathee, K., & Narasimhan, G. (2016). Metagenomics, metatranscriptomics, and metabolomics approaches for microbiome analysis. *Evolutionary Bioinformatics, 12*(S1), 5–16.

Araujo, F. A., Barh, D., Silva, A., Guimaraes, L., & Ramos, R. T. J. (2018). GO FEAT: A rapid web-based functional annotation tool for genomic and transcriptomic data. *Scientific Reports, 8*, 1794.

Bikel, S., Valdez-Lara, A., Cornejo-Granados, F., Rico, K., Canizales-Quinteros, S., Soberón, X., Pozo-Yauner, L. D., & Ochoa-Leyva, A. (2015). Combining metagenomics, metatranscriptomics and viromics to explore novel microbial interactions: Towards a systems-level understanding of human microbiome. *Computational and Structural Biotechnology Journal, 13*, 390–401.

Byron, S. A., Van Keuren-Jensen, K. R., Engelthaler, D. M., Carpten, J. D., & Craig, D. W. (2016). Translating RNA sequencing into clinical diagnostics: Opportunities and challenges. *Nature Reviews. Genetics, 17*, 257–271.

Cai, Z., Wang, L., Song, X., Tagore, S., Li, X., Wang, H., et al. (2018). Adaptive transcriptome profiling of subterranean Zokor, *Myospalax baileyi*, to high- altitude stresses in Tibet. *Scientific Reports, 8*, 4671.

Czypionka, T., Krugman, T., Altmuller, J., Blaustein, L., Steinfartz, S., Templeton, A. R., & Nolte, A. W. (2015). Ecological transcriptomics – A non-lethal sampling approach for endangered fire salamanders. *Methods in Ecology and Evolution, 6*, 1417–1425.

Derome, N., Duchesne, P., & Bernatchez, L. (2006). Parallelism in gene transcription among sympatric lake whitefish ecotypes (*Coregonus clupeaformis Mitchill*). *Molecular Ecology, 15*, 1239–1250.

Dhanasekaran, S. M., Barrette, T. R., Ghosh, D., Shah, R., Varambally, S., Kurachi, K., et al. (2001). Delineation of prognostic biomarkers in prostate cancer. *Nature, 412*, 822–826.

Durmuş, S., Cakir, T., Ozgur, A., & Guthke, R. (2015). A review on computational systems biology of pathogen-host interactions. *Frontiers in Microbiology, 6*, 235.

Eng, C. H. L., Shah, S., Thomassie, J., & Cai, L. (2017). Profiling the transcriptome with RNA SPOTs. *Nature Methods, 14*(12), 1153–1155.

Garcia-Sanchez, S., Aubert, S., Iraqui, I., Janbon, G., Ghigo, J. M., & d'Enfert, C. (2004). Candida albicans biofilms: A developmental state associated with specific and stable gene expression patterns. *Eukaryotic Cell, 3*, 536–545.

Garg, R., Shankar, R., Thakkar, B., Kudapa, H., Krishnamurthy, L., Mantri, N., et al. (2016). Transcriptome analyses reveal genotype- and developmental stage-specific molecular responses to drought and salinity stresses in chickpea. *Scientific Reports, 6*, 19228.

Gilbert, J. A., & Hughes, M. (2011). Gene expression profiling: metatranscriptomics. *Methods in Molecular Biology, 733*, 195–205.

Golub, T. R., Slonim, D. K., Tamayo, P., Huard, C., Gaasenbeek, M., Mesirov, J. P., et al. (1999). Molecular classification of cancer: Class discovery and class prediction by gene expression monitoring. *Science, 286*, 531–537.

Govindarajan, R., Duraiyan, J., & Palanisamy, M. (2012). Microarray and its applications. *Journal of Pharmacy and Bioallied Sciences, 4*(6), 310–312.

Hobbs, M., Pavasovic, A., King, A. G., Prentis, P. J., Eldridge, M. D., Chen, Z., et al. (2014). A transcriptome resource for the koala (*Phascolarctos cinereus*): Insights into koala retrovirus transcription and sequence diversity. *BMC Genomics, 15*, 786.

Hoheisel, J. D. (2006). Microarray technology: Beyond transcript profiling and genotype analysis. *Nature Reviews, 7*, 200–210.

Howe, G. T., Yu, J., Knaus, B., Cronn, R., Kolpak, S., Dolan, P., et al. (2013). A SNP resource for Douglas-fir: De novo transcriptome assembly and SNP detection and validation. *BMC Genomics, 14*, 137.

Hrdlickova, R., Toloue, M., & Tian, B. (2017). RNA-Seq methods for transcriptome analysis. *Wiley Interdiscip Rev RNA, 8*(1), e1364.

Jiang, Y., Xiong, X., Danska, J., & Parkinson, J. (2016). Metatranscriptomic analysis of diverse microbial communities reveals core metabolic pathways and microbiome specific functionality. *Microbiome, 4*, 2.

Kanter, I., & Kalisky, T. (2015). Single cell transcriptomics: Methods and applications. *Frontiers in Oncology, 5*, 53.

Li, B., & Dewey, C. N. (2011). Rsem: Accurate transcript quantification from RNA-seq data with or without a reference genome. *BMC Bioinformatics, 12*(1), 323.

Lin, Z., Chen, M., Dong, X., Zheng, X., Huang, H., Xu, X., & Chen, J. (2017). Transcriptome profiling of *Galaxea fascicularis* and its endosymbiont *Symbiodinium* reveals chronic eutrophication tolerance pathways and metabolic mutualism between partners. *Scientific Reports, 7*, 42100.

Lowe, R., Shirley, N., Bleackley, M., Dolan, S., & Shafee, T. (2017). Transcriptomics technologies. *Plos Computational Biology, 13*, e1005457.

Macgregor, P. F., & Squire, J. A. (2002). Application of microarrays to the analysis of gene expression in cancer. *Clinical Chemistry, 48*(8), 1170–1177.

McGrath, L. L., Vollmer, S. V., Kaluziak, S. T., & Ayers, J. (2016). De novo transcriptome assembly for the lobster Homarus americanus and characterization of differential gene expression across nervous system tissues. *BMC Genomics, 17*, 63.

Moncada, R., Chiodin, M., Devlin, J. C., Baron, M., Hajdu, C. H., Simeone, D., & Yanai, I. (2018 Jan 1). Building a tumor atlas: integrating single-cell RNA-Seq data with spatial transcriptomics in pancreatic ductal adenocarcinoma. *bioRxiv*, 254375.

Moreno, J. C., Pauws, E., van Kampen, A. H., Jedličková, M., de Vijlder, J. J., & Ris-Stalpers, C. (2001). Cloning of tissue-specific genes using SAGE and a novel computational substraction approach. *Genomic, 75*, 70–76.

Salem, M., Paneru, B., Al-Tobasei, R., Abdouni, F., Thorgaard, G. H., Rexroad, C. E., & Yao, J. (2015). Transcriptome assembly, gene annotation and tissue gene expression atlas of the rainbow trout. *PLoS One, 10*(3), e0121778.

Sun, M., Li, Y. T., Liu, Y., Lee, S. C., & Wang, L. (2015). Transcriptome assembly and expression profiling of molecular responses to cadmium toxicity in hepatopancreas of the freshwater crab *Sinopotamon henanense. Scientific Reports, 6*, 19405.

Velculescu, V. E., Zhang, L., Vogelstein, B., & Kinzler, K. W. (1995). Serial analysis of gene expression. *Science, 270*(5235), 484–487.

Wang, Z., Gerstein, M., & Snyder, M. (2009). RNA-Seq: A revolutionary tool for transcriptomics. *Nature Reviews. Genetics, 10*(1), 57–63.

Westermann, A. J., Gorski, S. A., & Vogel, J. (2012). Dual RNA-seq of pathogen and host. *Nature Reviews. Microbiology, 10*, 618–630.

Westreich, S. T., et al. (2016). SAMSA: A comprehensive metatranscriptome analysis pipeline. *BMC Bioinformatics, 17*(1), 399.

Westreich, S. T., Treiber, M. L., Mills, D. A., Korf, I., & Lemay, D. G. (2018). SAMSA2: A standalone metatranscriptome analysis pipeline. *BMC Bioinformatics, 19*(1), 175.

Wolf, J. B. W. (2013). Principles of transcriptome analysis and gene expression quantification: An RNA-seq tutorial. *Molecular Ecology Resources, 13*, 559–572.

Yamamato, M., Wakatsuki, T., Hada, A., & Ryo, A. (2001). Use of serial analysis of gene expression (SAGE) technology. *Journal of Immunological Methods, 250*, 45–66.

Yee, J. C., Gerdtzen, Z. P., & Hu, W. S. (2008). Comparative transcriptome analysis to unveil genes affecting recombinant protein productivity in mammalian cells. *Biotechnology and Bioengineering, 102*, 246–263.

# Chapter 4
# Proteomics

**Candida Vaz and Vivek Tanavde**

**Abstract** Study of the complete proteins from a source and the techniques implied to study these proteins and their interactions is the fundamental of proteomics. The three-dimensional map of the proteins and their interactions delineates their importance and functioning in an organism. These studies are initiated at the protein level sometimes tracing back to their genes. The alternative splicing phenomenon in the eukaryotes selectively enriches the proteome diversity. Structural proteomics takes into consideration the three-dimensional structure of proteins helping in the structure-based rational drug designing procedure. On the other hand, the functional proteomics is largely focused on understanding the protein expression at the cellular level, protein modifications, protein interactions, signalling and disease mechanisms. The field has gained momentum with the advent of technology where different techniques such as X-ray, NMR, mass spectroscopy, HPLC and two-dimensional PAGE have resulted in the generation of enormous experimental data.

It is difficult to keep up with the colossal amount of experimental data generated through various protein detection methods. The analysis done through bioinformatics procedures involving algorithms, databases and pipelines for computational analysis enables faster and accurate analysis done over a couple of days. Through databases and resource portals, data management, storage and sharing have made it easier for researchers to obtain and collate data accelerating proteomics research.

This chapter is an effort to describe in detail the various dimensions of the proteomics studies covering the structural and functional aspects. A brief overview of all the techniques involved in studying the proteome is given facilitating the reader

C. Vaz
Bioinformatics Institute, Agency for Science Technology and Research (A*STAR),
Singapore, Singapore
e-mail: candidav@bii.a-star.edu.sg

V. Tanavde (✉)
Bioinformatics Institute, Agency for Science Technology and Research (A*STAR),
Singapore, Singapore

Division of Biological and Life Sciences, School of Arts and Sciences,
Ahmedabad University, Ahmedabad, India
e-mail: vivek.tanavde@ahduni.edu.in

© Springer Nature Singapore Pte Ltd. 2018
P. Arivaradarajan, G. Misra (eds.), *Omics Approaches, Technologies
And Applications*, https://doi.org/10.1007/978-981-13-2925-8_4

to incorporate the ideas in their research planning. The immense importance of the domain in biomarker discovery and elucidation of protein-protein/drug interactions.

Proteomics in combination with other complementary technologies like genomics and transcriptomics (a systems-level approach) has an enormous potential to answer several unanswered questions in biology.

**Keywords** Proteomics · Proteogenomics · Protein diversity · Drug discovery

## 4.1 Understanding Proteomics

Proteins are vital molecules that have direct involvement in cellular function. According to the central dogma of molecular biology, the DNA synthesizes the RNA that in turn synthesizes the protein. The DNA contains the blueprint on how to assemble a cell, but it is the proteins that ultimately serve as the building blocks.

"Proteome", a term coined by Marc Wilkins in 1994 (Wasinger et al. 1995), is the study of the entire range of proteins in a single cell. The term "proteomics" was coined in 1997 (James 1997) and refers to the large-scale study of proteomes that involves research and exploration of the proteomes pertaining to their structure, composition, function and activity patterns.

Following "genomics" and "transcriptomics" that comprise the study of the genome and transcriptome, respectively, "proteomics" is the next major step in the study of biomolecular systems. Molecular biology has provided substantial techniques for high-throughput nucleotide sequence analysis that needs to be reflected in the protein world. It is essential to make sense of the huge amount of sequence data being generated.

There exist two modes of proteomics, one that involves the study of only proteins as analysis of gene products and the other that is more inclusive and comprises a combination of protein analysis and genomics/transcriptomics. Different areas of studying proteins such as protein function, modification, interaction and localization are now grouped under the broad definition of proteomics.

However, proteomics is complex, and its complexity exists in the fact that an organism's genome remains more or less constant or static, whereas its proteome is highly dynamic, differing in context of cell and time. In response to both external and internal factors, proteins can be synthesized or degraded or undergo post-translational modifications. Therefore, studying a proteome is actually like taking a snapshot of the protein environment at a given time. A single genome can give rise to a large number of different proteomes depending on factors like cell cycle stage, growth, nutrition, pathological conditions and stress.

The following features need to be fulfilled by an ideal proteomics technology: high sensitivity, high throughput, the ability to differentiate modified proteins and the ability to quantitatively display and analyze all the proteins in a sample (Haynes and Yates 2000).

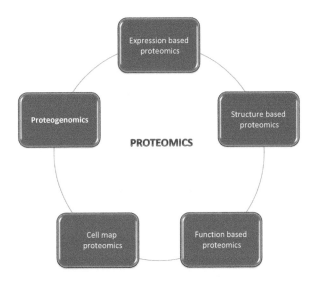

**Fig. 4.1** Different dimensions of proteomics

It has wide applications in the field of clinical and biomedical sciences, wherein alterations in the proteome of tissues or body fluids can be measured and correlated with diseases, disorders, treatment or hormone action. The different dimensions of proteomics are outlined in Fig. 4.1.

## 4.2   The Need for Proteomics

Genomics alone cannot provide answers to all the questions. The phenotype is largely governed by proteins and not by genes. It is difficult and almost impossible to unravel the mechanisms of diseases, disorders and ageing just by studying the genome.

As most of the drug targets are proteins, the interpretation of the genome at the protein level is invaluable, and that is what proteomics aims to achieve (Blackstock and Weir 1999).

RNA analysis was used earlier to determine the protein content, but it was found to be inaccurate, as it did not correlate well with protein expression (Abbott 1999; Anderson and Seilhamer 1997; Gygi et al. 1999a; Ideker et al. 2001a). The mRNA is not directly correlated to the proteins in a cell as following the transcription step (post-transcription) there are a series of events in the form of alternative splicing and mRNA editing (Newman 1998) that can create several different forms of a protein (isoforms). In addition to this, the proteins themselves undergo post-translational modifications, and there are almost up to 200 types of such protein modifications (Krishna and Wold 1993). Proteolysis (Kirschner 1999) and compartmentalization (Colledge and Scott 1999) also regulate proteins and contribute to protein diversity. The different steps followed for performing a proteomics study are outlined in Fig. 4.2.

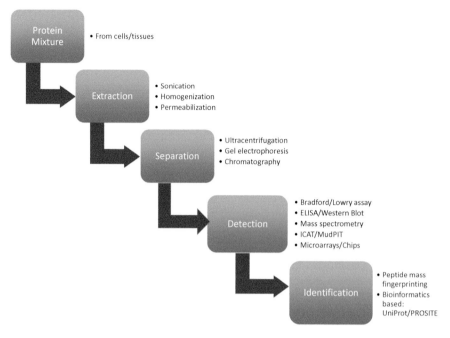

**Fig. 4.2** Common workflow for a proteomic study

## 4.3 Different Categories of Proteomics

### 4.3.1 Protein Expression-Based Proteomics

This category involves the quantification of the expression of the proteome followed by comparison of the protein expression among samples that differ by a factor being studied. The factor could be a disease, a drug treatment or an environmental effect. The differentially expressed proteins identified are highly valuable as disease-/condition-specific proteins or as potential drug targets or as diagnostic markers.

### 4.3.2 Structure-Based Proteomics

This category involves large-scale analysis of protein structures. Deducing the protein structures helps in the identification and assignment of functions to newly discovered genes through a process called annotation. Structural analysis of proteins is essential for determining the protein interactions and for drug-binding studies. Structure-based proteomics is done mainly through X-ray crystallography and NMR spectroscopy.

### 4.3.3 Function-Based Proteomics

This is a broader category and involves several proteomics approaches ranging from discovery and identification of novel proteins/protein complexes, study and characterization of proteins for elucidating protein signalling, interaction and disease mechanisms.

### 4.3.4 Cell Map Proteomics

This category involves studies aiming at identifying the structure of protein complexes or the proteins present inside the cell or a particular organelle creating a "three-dimensional cell map" (Blackstock and Weir 1999). These studies determine where the proteins are located and their interactions with other proteins. The importance of such studies is their ability to determine the overall architecture of the cells and to determine how the expression of certain proteins gives the cell its unique characteristics. The identification of the nuclear pore complex is a good example of this category of proteomics (Rout et al. 2000).

### 4.3.5 Proteogenomics

This category involves the use of proteomics for improving gene annotations and genomics. A parallel and combined analysis of the proteome and the genome accelerates the discovery of regulatory mechanisms such as post-transcriptional and post-translational modifications (Gupta et al. 2007). This approach helps in analyzing and comparing multiple genomes (Gupta et al. 2008).

## 4.4 Protein Detection Methods

### 4.4.1 Non-specific Detection Methods that Rely on the Absorbance Property

#### 4.4.1.1 Bradford Protein Assay

It is a spectroscopic analytical procedure that was developed by Marion M. Bradford in 1976 (Bradford 1976). It is based on the absorbance shift of the Coomassie brilliant blue G-250 dye that exists in three forms: anionic (blue), neutral (green) and cationic (red). Under acidic conditions and when not bound to a protein, the dye is in a protonated red cationic form with maximum absorbance at 470 nm. When binding to a protein occurs, it is converted to an unprotonated blue form with absorbance at 595 nm. The binding of the protein causes the stabilization of the blue form of the dye, and by measuring the amount of the complex present (through absorbance readings) in the solution, the protein concentration can be estimated. The increase of absorbance from 465 nm to 595 nm is equivalent to the amount of bound dye that is in proportion to the protein present in the sample. The advantage of this technique is that it is simple, fast and sensitive. The disadvantages of this technique are mainly related to the conditions and detergents that can interfere with the dye binding to the protein efficiently.

This method is highly dependent on protein sequence and if the protein does not contain a sufficient number of aromatic residues efficient binding of the dye won't occur. Moreover this method relies on the comparison of the absorbance of the protein to that of a standard protein, so if a protein differs from the standard protein in the way it reacts to the dye, the estimation of its concentration will not be accurate.

### 4.4.1.2    Lowry Protein Assay

This assay used for determining the total level of protein in a solution was developed by Oliver H. Lowry (1951). It is a variant of the Biuret test, which can assess the concentration of proteins in a sample by detection of the peptide bonds that leads to a change in colour (purple) that can be measured by a colorimetric test at 550 nm.

The Lowry method involves the reaction of copper ions produced by the oxidation of the peptide bonds with the Folin-Ciocalteu reagent. The reduction of the Folin-Ciocalteu reagent and oxidation of the aromatic residues are the crux of the reaction mechanism that results in an intense blue molecule called the heteropolymolybdenum blue. The concentration of the protein in the sample can be determined from the concentration of the aromatic residues that reduce the reagent. The concentration of the reduced reagent can be measured by colorimetric techniques (absorbance at 660 nm).

## 4.4.2    Specific Detection Methods Using Antibodies

Antibodies are used for quantitative/semi-quantitative detection of specific proteins, a method commonly called as immunoassay. It is sensitive, specific and cost-effective.

Production of antibodies using monoclonal techniques made the large-scale utilization of this method possible (Kohler and Milstein 2005).

The normal class of antibody used in an immunoassay is an IgG molecule, and the area of the protein target that is recognized is called an epitope made up of 10–15 amino acids. An epitope may be continuous when it is made up of amino acids that are sequential in primary sequence of a protein or it may be discontinuous when the amino acids are distant in the primary sequence but are brought together by the secondary and tertiary structure of the protein. Antibodies specific to a modification, like phospho-specific antibodies, can detect modified proteins that are tyrosine phosphorylated.

### 4.4.2.1 Enzyme-Linked Immunosorbent Assay

This method makes use of specific antibodies that are linked to an enzyme (detection antibodies) (Lequin 2005). The sample comprising of unknown amount of proteins/antigens is attached to a surface and exposed to these enzyme-linked antibodies. The sample with the antigens can be immobilized on the solid support like polystyrene microtiter plate. After the immobilization of the antigen, the detection antibodies are added that bind to the antigens.

After the binding of the antibodies to the antigens, the substrate of the particular enzyme is added. The addition of the substrate creates a reaction that produces a detectable signal like a colour change, which is indicative of the quantity of antigen in the sample. The detection of the intensity of the transmitted light is done by a spectrophotometer.

### 4.4.2.2 Western Blot

It is a widely used method to detect specific proteins in a sample. This procedure makes use of antibodies that can react with a specific protein target (Burnette 1981; Alwine et al. 1977). The sample is first subjected to protein denaturation followed by gel electrophoresis to separate the proteins. SDS is used as a buffer that maintains the polypeptides in a denatured state and also confers upon them a uniform negative charge. This type of electrophoresis is called SDS-PAGE (Laemmli 1970). After the electrophoretic separation of the proteins, they are transferred onto a nitrocellulose membrane where they are exposed to antibodies specific to the target proteins. The next step is to use a secondary antibody that recognizes the first antibody. The secondary antibody can allow the visualization of the protein through staining, immunofluorescence and radioactivity.

## 4.4.3   Specific Detection Methods Without the Use of Antibodies

### 4.4.3.1   2D Electrophoresis (2DE)

2D electrophoresis using isoelectric focusing/sodium dodecyl sulphate gel electrophoresis (IEF/SDS-PAGE) is the most commonly used method in proteomics (O'Farrell 1975; Klose 1975). Proteins first separated by isoelectric focusing are further resolved by SDS-PAGE. Staining or autoradiography of these separated proteins produces a 2D array that helps visualize them. The identification of individual proteins from such gels has been carried out using techniques such as co-migration with known proteins, immunoblotting, N-terminal sequencing or internal peptide sequencing.

2D electrophoresis is time-consuming, nonquantitative, inaccurate, of limited range and non-suitable for hydrophobic proteins. SDS is incompatible with IEF, making analysis of hydrophobic proteins especially problematic in 2D gels. Significant progress has been made to overcome this limitation by the development of detergents that have better solubilizing power and by the selective use of organic solvents that promote the solubility of hydrophobic proteins. Another disadvantage is that 2DE is able to identify only the most abundant and long-lived proteins that have optimal codons. Proteins that are less abundant and/or with codon bias index of less than 0.1 are unlikely to be visualized without prior enrichment methods. Despite these disadvantages, 2DE is used for its ability to visualize a large number of proteins simultaneously for differential display experiments where the proteins could be subjected to gene knockouts, cell differentiation, potential drug treatments or changes in growth/nutrition conditions.

In identification of the pathways, the differentially expressed proteins are involved in helps in paving the path for future research.

### 4.4.3.2   Mass Spectrometry

This is an analytical technique that measures the masses within a sample using the process of ionization of the sample and then sorting of ions based on their mass-to-charge ratio (Price 1991).

A mass spectrum is generated that consists of a plot of the ion signal as a function of mass-to-charge ratio. These spectrums are used to elucidate the chemical structures and determine the masses of peptides. A typical MS procedure involves the following steps: (i) ionization of the sample by bombarding it with electrons, (ii) separation of the charged ions of the sample according to their mass-to-charge ratios by subjecting them to an electric/magnetic field, (iii) detection of ions by a mechanism capable of detecting charged particles, (iv) display of results as spectra of the relative abundance of the detected ions as a function of the mass-to-charge ratio and (v) identification of the atoms/molecules in the sample by correlation of the known masses to the identified masses or through an already known characteristic fragmentation pattern.

The two main techniques for gas-phase ionization of large, polar and highly charged molecules are electrospray ionization (ESI) (Ho et al. 2003) and matrix-assisted laser desorption ionization (MALDI) (Hillenkamp et al. 1991). In ESI, ions are formed from a liquid solution at atmospheric pressure. In MALDI, a laser pulse causes sublimation of the sample out of a dry crystalline matrix.

The devices that are used to bring the samples into gas phase and ionize them are called "sources". Commonly used four types of mass analyzers for protein studies differ in their design: the ion trap (IT), quadrupole (Q), time of flight (TOF) and Fourier-transform ion cyclotron (FT-ICR). Two analyzers can be placed in tandem to create a two-stage mass spectrometry called "tandem MS" or "MS/MS".

Coupling of different analyzers with different sources creates a large number of instrumental configurations. Two most commonly used configurations that have produced most of the published proteomic data so far are the ion traps coupled to ESI sources (ESI-IT) and the TOFs coupled to MALDI source (MALDI-TOF).

MS experiments for protein identification are mostly peptide-based analysis and comprise the following steps:

(i) Obtaining the proteins of interest from cells/tissues through mainly biochemical fractionation based on chromatographic techniques.

(ii) Fractionation of the protein sample through 1D gel or by 2D-PAGE followed by excision and in-gel digestion of the gel band (1D gel) or spot (2D gel) using proteases or chemicals or by "gel-free" fractionation of the samples that is done by using two-dimensional fractionation of the peptide mixtures that involves reduced and digested peptide mixture fractionated through a strong cationic exchange column (SCX) and further separated on a reverse phase (RP) column. This technique is known as multidimensional protein identification technology (MudPIT) (Schirmer et al. 2003). Peptide mixtures generated by digesting multiple proteins have to be first separated by high-pressure liquid chromatography (HPLC) before being introduced to MS. Elution of the peptides from the RP columns is done using increasing concentrations of organic solvents. An LC-ESI-MS technique consists of a chromatographic column coupled online with a MS fitted to an ESI source.

(iii) Analysis of the peptide mixture by MS.

(iv) Identification of the proteins by matching a list of experimental peptide masses with the theoretically calculated masses using computer-generated list formed from the simulated digestion of a protein database using the same enzyme through an approach called "peptide mass fingerprinting" (PMF) (Pappin et al. 1993).

### 4.4.3.3   Isotope-Coded Affinity Tag Peptide Labelling (ICAT)

This method consists of a twofold approach of sequence identification and accurate quantification of the proteins. It uses a chemical reagent called isotope-coded affinity tag (ICATs) in combination with tandem mass spectroscopy (Gygi et al. 1999b). The ICAT reagent comprises a biotin affinity tag joined by a spacer domain to a thiol-specific reactive group, which exists in two forms: regular and isotopically heavy. The method involves four steps: (i) derivatization of the reduced protein mixtures representing two different states with the isotopically light and heavy versions of the ICAT reagent, (ii) proteolytic digestion of the labelled samples to produce peptide fragments, (iii) isolation of the tagged cysteine containing peptide fragments by avidin affinity chromatography and (iv) separation and analysis of the isolated tagged peptides by microcapillary tandem mass spectrometry that provides the identification of the peptides by fragmentation and relative quantification of the labelled pairs through comparison of the signal intensities in MS mode

The advantages of this method are that it is non-time-consuming and scaleable and enables the analysis of low abundant proteins. Since this method is based on stable isotopic labelling, it does not require the use of radioactivity or metabolite labelling. The method provides accurate quantification each peptide identified. The disadvantages of this method are the proteins must contain cysteine that must be flanked by protease cleavage sites; in case of small peptide, the ICAT tag may interfere with the peptide ionization and complicate the mass spectral results. However, these disadvantages can be overcome by designing different reagents for increased specificity, using a smaller tag and by using different types of proteases.

### 4.4.3.4 Multidimensional Protein Identification Technique (MudPIT)

This method combines the usage of multidimensional liquid chromatography and tandem mass spectroscopy (Schirmer et al. 2003). It comprises the following steps: (i) digestion of the denatured and reduced protein mixture to produce peptide fragments; (ii) loading of the mixture onto a microcapillary column containing SCX resin upstream of RPC resin that elutes directly into a tandem mass spectrometer; (iii) displacement of the absorbed peptides from the SCX column onto the RPC column using a salt gradient, which causes the peptide to be retained on the RPC column; (iv) elution of the peptides from the RPC column using an acetonitrile gradient; (v) analysis by MS/MS; (vi) repetition of the process using increased concentration of salt to displace additional fraction from the SCX column in an iterative manner involving 10–20 steps; and (vii) analysis of the MS/MS data from all of the fractions by database searching followed by combination to give the general picture of the proteins present in the sample.

There are several advantages of this technique, viz. non-time-consuming, amenable to full automation, increases the number of peptides being identified from complex mixtures, has a very wide dynamic range and no solubility problems associated with 2DE. The drawbacks of this method are mainly related to the sheer volume of data generated by a MudPIT experiment consisting of 10–20 cycles of reversed phase chromatography that leads to problems associated with computing power. This approach is also only useful for organisms where complete genome sequence data is available.

### 4.4.3.5 Protein Microarrays/Protein Chips

Protein microarrays are used extensively for protein expression profiling as well as for detecting protein-protein interactions (Melton 2004). This technology uses thousands of protein detecting features for probing biological samples. Multiple protein types can be arrayed for studying protein-DNA, protein-protein and protein-ligand interactions.

A functional proteomic array typically contains the entire complement of proteins from an organism. However, protein chips are difficult to implement as proteins are less stable and more dynamic, and it is difficult to maintain their structural integrity on the glass slide.

## 4.5 Bioinformatics in Proteomics

It is difficult to keep up with the vast amount of data generated through the various protein detection methods, and it is tedious to analyze this data manually. To accelerate proteomics research, the need for collaborating with computational scientists in particular the bioinformaticians is imperative to create programmes/algorithms, databases and pipelines for computational analysis of the data. The analysis done through bioinformatics procedures enables faster and accurate analysis done over a couple of days as compared to several weeks and months if analyzed manually. Moreover, data storage, management and sharing are done through databases and resource portals that have made it easier for a researcher to collate information or obtain data. The contribution of bioinformatics in proteomics is highlighted in Fig. 4.3.

The example of such a resource portal for proteomics is the **ExPASy (Expert Protein Analysis System)** that is operated by the **Swiss Institute of Bioinformatics (SIB)**. It is a single web portal that provides access to several databases, resources and tools developed by several SIB institutes and external organizations (Gasteiger et al. 2003). Another resource portal is **UniProt** that provides comprehensive, high-quality and freely accessible resource for protein structure and functional information (UniProt 2015).

**The PDB (Protein Data Bank)** is a resource that aims at providing a structural view of biology by furnishing information about the 3D shapes of proteins, nucleic acids and complex assemblies (Berman 2008).

### 4.5.1 Bioinformatics for Protein Identification

The protein detection methods like microarray and mass spectrometry detect peptide fragments only and are not capable of identifying the entire protein in the sample. There are now several programmes that help in identification of the proteins by matching the peptide fragments to the known proteins in databases. This peptide to protein matching is carried out by algorithms that perform alignments with known proteins in databases such as UniProt (UniProt 2015) and PROSITE (de Castro et al. 2006).

**Fig. 4.3** Use of bioinformatics for proteomics

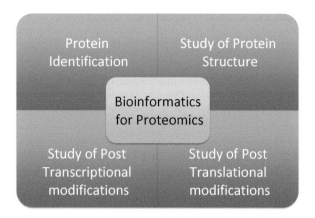

## 4.5.2   Bioinformatics for Protein Structural Studies

It is essential to understand the protein structure to be able to interpret its interactions and functions. The 3D structure of proteins could be earlier, only be determined using X-ray crystallography (Ilari and Savino 2008) and NMR spectroscopy (Wuthrich 2001). Recently, cryo-electron microscopy (Costa et al. 2017) has become the leading technique overcoming the limitations of crystallization and conformational ambiguity related to X-ray and NMR techniques, respectively.

An array of programmes/tools that can predict and model the structure of proteins are now available. These programmes/tools use the chemical properties of amino acids and the properties of known proteins to predict and model the structure of unknown proteins (Table 4.1).

**Table 4.1**   Tools for structural analysis (Ghorbani et al. 2016)

| Tools | Function | URL |
|---|---|---|
| **Cn3D** | Facilitates viewing of 3D structures from NCBI's Entrez Structure database. It can simultaneously display sequence, structure, and alignment | https://www.ncbi.nlm.nih.gov/Structure/CN3D/cn3d.shtml |
| **DeepView Swiss-PDB viewer** | Facilitates the analysis of structure, alignments, homology modeling, and searching for functional sites | https://spdbv.vital-it.ch |
| **RasMol and OpenRasMol** | Facilitates molecular graphics visualization | http://www.openrasmol.org |
| **Mage and Kinemage** | Facilitates the presentation of scientific illustrations as an interactive computer display | http://kinemage.biochem.duke.edu/kinemage/magepage.php#defined |

(continued)

**Table 4.1** (continued)

| Tools | Function | URL |
|---|---|---|
| **Vector Alignment Search Tool (VAST)** | Facilitates identification of similar protein 3D structures by purely geometric criteria and identification of distant homologs that cannot be determined through sequence comparison | https://structure.ncbi.nlm.nih.gov/Structure/VAST/vast.shtml |
| **DALI** | Facilitates comparison of 3D protein structures by comparing the coordinates of a query protein structure against those in the PDB | http://ekhidna2.biocenter.helsinki.fi/dali/ |
| **Conserved Domain Database (CDD)** | This database is a collection of multiple sequence alignment models for domains of proteins. Facilitates fast identification of conserved domains in protein sequences | https://www.ncbi.nlm.nih.gov/Structure/cdd/cdd.shtml |
| **Conserved Domain Architecture Retrieval Tool (CDART)** | Shows the functional domains found in the query protein and lists proteins having the similar domain architecture | https://www.ncbi.nlm.nih.gov/Structure/lexington/lexington.cgi |
| **LIGPLOT** | Generates schematic 2D diagrams of ligand-protein interaction | https://www.ebi.ac.uk/thornton-srv/software/LIGPLOT/ |
| **Structural Classification of Proteins (SCOP)** | Provides comprehensive information of the structural and evolutionary relationships between proteins whose structures are known | http://scop.mrc-lmb.cam.ac.uk/scop/ |
| **CATH** | Provides classification of protein structures. Helps in predicting protein function through structure and sequence | http://www.cathdb.info |

## 4.5.3 Bioinformatics for Studying Post-transcriptional Modifications

Though mRNA/gene expression does not correlate well with the protein expression, it is still possible to determine translationally regulated genes. This can be achieved by using integrated transcriptome and translational state profiling. Differentially translated genes can be obtained using total and fractionated RNA microarrays. Fractionation of the RNA is done through a sucrose gradient. Sucrose gradient fractionation can be used to separate actively translated mRNAs that are associated with multiple ribosomes (polysomes) and the inactive mRNAs. As the number of ribosomes on a transcript correlates with the rate of synthesis of the protein encoded by it, this creates an operational distinction between the highly translated and poorly translated mRNA. This approach was used for identification of translationally regulated genes during embryonic stem cell differentiation (Sampath et al. 2011) (Table 4.2).

**Table 4.2** Tools for detection of post-translational modifications (PTM) (Audagnotto and Dal Peraro 2017)

| Tools | PTM type | URL |
|---|---|---|
| **NetPhos3.1** | Phosphorylation | http://www.cbs.dtu.dk/services/NetPhos/ |
| **PhosphoELM** | Phosphorylation | http://phospho.elm.eu.org |
| **GlycoMod** | Glycosylation | https://web.expasy.org/glycomod/ |
| **NetOGlyc** | Glycosylation | http://www.cbs.dtu.dk/services/NetOGlyc/ |
| **iSNO-PseAAC** | Methylation | http://app.aporc.org/iSNO-PseAAC/ |
| **MethK** | Methylation | http://csb.cse.yzu.edu.tw/MethK/ |
| **PAIL** | N-acetylation | http://bdmpail.biocuckoo.org/prediction.php |
| **N-Ace** | N-acetylation | http://n-ace.mbc.nctu.edu.tw |
| **UbPred** | Ubiquitylation | http://www.ubpred.org |
| **UbiNet** | Ubiquitylation | http://140.138.144.145/%7Eubinet/index.php |

### 4.5.4   Bioinformatics for Studying Post-translational Modifications

Since post-translational modifications can affect the protein structure, the computational analysis of these modifications is necessary and needs to be included in any protein study. New pipelines and tools are now being created for analyzing post-translational modifications (Margreitter et al. 2013).

## 4.6   Challenges of Proteomics

One of the major problems associated with proteomics is the low abundance proteins. Such low expressing proteins are difficult to detect in the analysis of crude cell lysates without sophisticated purification methods. Many important classes of proteins such as transcription factors, kinases and regulatory proteins are low expressing proteins. There is also no PCR for proteins as compared to DNA, and hence the analysis of low expressing proteins is difficult.

It is also difficult to study proteins on a scale comparable to genomic analysis. Most of the methods used for proteomics are not high-throughput. Although MS is widely used for protein identification and analysis, data procurement and analysis is still time-consuming. The quality of the data is also at a risk of being sacrificed. MS/MS can provide higher-quality data, but the data interpretation is highly time-consuming. This warrants the need for better computational algorithms for increasing the accuracy of data interpretation without any manual intervention.

It is difficult to measure proteins accurately and quantitatively. RNA measurements are more precise, and their sensitivity is down to a single transcript. Single molecule protein measurements are too difficult and cumbersome. Moreover, variations at the protein level are higher than the RNA level, which makes it difficult to classify samples based on protein expression alone.

Combining genomics, transcriptomics and proteomics will give us a better overview of what is happening inside the cell (Ideker et al. 2001b; Ghaemmaghami et al. 2003). While this concept is attractive, there are numerous challenges in data analysis that need to be overcome before this becomes a reality.

This systems-level approach has been attempted for relatively simple organisms like yeast (Ideker et al. 2001a). Ideker and Hood tried a systems approach to understand yeast. They defined all genes in the yeast genome. They also created subsets of genes involved in galactose pathway and built model of this pathway using the following steps: (i) perturb each pathway using genetic and environmental tools, (ii) use microarray to catalog all the changes in the transcriptome and ICAT to measure changes in the proteome in all these perturbations and (iii) refine the model by fitting the experimental data to the existing model.

There are also some problems associated with this systems approach, namely: (i) different platforms behave differently making it impossible to compare datasets of transcriptome and proteome, (ii) normalization of data is a huge challenge as it is very difficult to normalize data across different platforms, (iii) dealing with different file formats in each measurement and harmonizing these and (iv) annotation varies across DNA, RNA and proteins. For example, there exist certain genes for which protein products are unknown.

Despite these challenges, proteomics combined with other complementary technologies like genomics and transcriptomics has an enormous potential to answer several unanswered questions in biology.

# References

Abbott, A. (1999). A post-genomic challenge: Learning to read patterns of protein synthesis. *Nature, 402*(6763), 715–720.

Alwine, J. C., Kemp, D. J., & Stark, G. R. (1977). Method for detection of specific RNAs in agarose gels by transfer to diazobenzyloxymethyl-paper and hybridization with DNA probes. *Proceedings of the National Academy of Sciences of the United States of America, 74*(12), 5350–5354.

Anderson, L., & Seilhamer, J. (1997). A comparison of selected mRNA and protein abundances in human liver. *Electrophoresis, 18*(3–4), 533–537.

Audagnotto M., & Dal Peraro M. (2017 Mar 31). Protein post-translational modifications: In silico prediction tools and molecular modeling. *Computational and Structural Biotechnology Journal, 15*:307–319. doi: https://doi.org/10.1016/j.csbj.2017.03.004. eCollection 2017. https://www.ncbi.nlm.nih.gov/pubmed/28458782

Berman, H. M. (2008). The Protein Data Bank: A historical perspective. *Acta Crystallographica A, 64*(Pt 1), 88–95.

Blackstock, W. P., & Weir, M. P. (1999). Proteomics: Quantitative and physical mapping of cellular proteins. *Trends in Biotechnology, 17*(3), 121–127.

Bradford, M. M. (1976). A rapid and sensitive method for the quantitation of microgram quantities of protein utilizing the principle of protein-dye binding. *Analytical Biochemistry, 72*, 248–254.

Burnette, W. N. (1981). "Western blotting": Electrophoretic transfer of proteins from sodium dodecyl sulfate--polyacrylamide gels to unmodified nitrocellulose and radiographic detection with antibody and radioiodinated protein A. *Analytical Biochemistry, 112*(2), 195–203.

Colledge, M., & Scott, J. D. (1999). AKAPs: From structure to function. *Trends in Cell Biology, 9*(6), 216–221.

Costa, T. R. D., Ignatiou, A., & Orlova, E. V. (2017). Structural analysis of protein complexes by cryo electron microscopy. *Methods in Molecular Biology, 1615*, 377–413.

de Castro, E., et al. (2006). ScanProsite: Detection of PROSITE signature matches and ProRule-associated functional and structural residues in proteins. *Nucleic Acids Research, 34*(Web Server issue), W362–W365.

Gasteiger, E., et al. (2003). ExPASy: The proteomics server for in-depth protein knowledge and analysis. *Nucleic Acids Research, 31*(13), 3784–3788.

Ghaemmaghami, S., et al. (2003). Global analysis of protein expression in yeast. *Nature, 425*(6959), 737–741.

Ghorbani, M., Ghorbani, F., & Karimi, H. (2016). Bioinformatics tools for protein analysis. *IJSRST, 2*(4). https://www.researchgate.net/publication/305720975_Bioinformatics_Tools_for_Protein_Analysis.

Gupta, N., et al. (2007). Whole proteome analysis of post-translational modifications: Applications of mass-spectrometry for proteogenomic annotation. *Genome Research, 17*(9), 1362–1377.

Gupta, N., et al. (2008). Comparative proteogenomics: Combining mass spectrometry and comparative genomics to analyze multiple genomes. *Genome Research, 18*(7), 1133–1142.

Gygi, S. P., et al. (1999a). Correlation between protein and mRNA abundance in yeast. *Molecular and Cellular Biology, 19*(3), 1720–1730.

Gygi, S. P., et al. (1999b). Quantitative analysis of complex protein mixtures using isotope-coded affinity tags. *Nature Biotechnology, 17*(10), 994–999.

Haynes, P. A., & Yates, J. R., 3rd. (2000). Proteome profiling-pitfalls and progress. *Yeast, 17*(2), 81–87.

Hillenkamp, F., et al. (1991). Matrix-assisted laser desorption/ionization mass spectrometry of biopolymers. *Analytical Chemistry, 63*(24), 1193A–1203A.

Ho, C. S., et al. (2003). Electrospray ionisation mass spectrometry: Principles and clinical applications. *Clinical Biochemist Review, 24*(1), 3–12.

Ideker, T., et al. (2001a). Integrated genomic and proteomic analyses of a systematically perturbed metabolic network. *Science, 292*(5518), 929–934.

Ideker, T., Galitski, T., & Hood, L. (2001b). A new approach to decoding life: Systems biology. *Annual Review of Genomics and Human Genetics, 2*, 343–372.

Ilari, A., & Savino, C. (2008). Protein structure determination by x-ray crystallography. *Methods in Molecular Biology, 452*, 63–87.

James, P. (1997). Protein identification in the post-genome era: The rapid rise of proteomics. *Quarterly Reviews of Biophysics, 30*(4), 279–331.

Kirschner, M. (1999). Intracellular proteolysis. *Trends in Cell Biology, 9*(12), M42–M45.

Klose, J. (1975). Protein mapping by combined isoelectric focusing and electrophoresis of mouse tissues. A novel approach to testing for induced point mutations in mammals. *Humangenetik, 26*(3), 231–243.

Kohler, G., & Milstein, C. (2005). Continuous cultures of fused cells secreting antibody of predefined specificity. 1975. *Journal of Immunology, 174*(5), 2453–2455.

Krishna, R. G., & Wold, F. (1993). Post-translational modification of proteins. *Advances in Enzymology and Related Areas of Molecular Biology, 67*, 265–298.

Laemmli, U. K. (1970). Cleavage of structural proteins during the assembly of the head of bacteriophage T4. *Nature, 227*(5259), 680–685.

Lequin, R. M. (2005). Enzyme immunoassay (EIA)/enzyme-linked immunosorbent assay (ELISA). *Clinical Chemistry, 51*(12), 2415–2418.

Lowry, O. H., et al. (1951). Protein measurement with the Folin phenol reagent. *Journal of Biological Chemistry, 193*(1), 265–275.

Margreitter, C., Petrov, D., & Zagrovic, B. (2013). Vienna-PTM web server: A toolkit for MD simulations of protein post-translational modifications. *Nucleic Acids Res, 41*(Web Server issue), W422–W426.

Melton, L. (2004). Protein arrays: Proteomics in multiplex. *Nature, 429*(6987), 101–107.

Newman, A. (1998). RNA splicing. *Current Biology, 8*(25), R903–R905.

O'Farrell, P. H. (1975). High resolution two-dimensional electrophoresis of proteins. *Journal of Biological Chemistry, 250*(10), 4007–4021.

Pappin, D. J., Hojrup, P., & Bleasby, A. J. (1993). Rapid identification of proteins by peptide-mass fingerprinting. *Current Biology, 3*(6), 327–332.

Price, P. (1991). Standard definitions of terms relating to mass spectrometry: A report from the committee on measurements and standards of the American society for mass spectrometry. *Journal of the American Society for Mass Spectrometry, 2*(4), 336–348.

Rout, M. P., et al. (2000). The yeast nuclear pore complex: Composition, architecture, and transport mechanism. *Journal of Cell Biology, 148*(4), 635–651.

Sampath, P., Lee, Q. Y., & Tanavde, V. (2011). Identifying translationally regulated genes during stem cell differentiation. *Current Protocols in Stem Cell Biology, Chapter 1*, Unit1B 8.

Schirmer, E. C., Yates, J. R., 3rd, & Gerace, L. (2003). MudPIT: A powerful proteomics tool for discovery. *Discovery Medicine, 3*(18), 38–39.

UniProt, C. (2015). UniProt: A hub for protein information. *Nucleic Acids Research, 43*(Database issue), D204–D212.

Wasinger, V. C., et al. (1995). Progress with gene-product mapping of the Mollicutes: Mycoplasma genitalium. *Electrophoresis, 16*(7), 1090–1094.

Wuthrich, K. (2001). The way to NMR structures of proteins. *Nature Structural & Molecular Biology, 8*(11), 923–925.

# Chapter 5
# Metabolomics

**Abhishek Sengupta and Priyanka Narad**

**Abstract** Metabolomics is formally defined as the high-throughput study of metabolites which serve as an integral part of the metabolism. With the advent of Human Genome Project, a plethora of data repositories have evolved generating huge amounts of 'omics' data. These data can be classified as genomics, proteomics, transcriptomics and metabolomics. However, amongst these metabolomics directly emulates the biochemical activity of the organism and thus best describes the molecular phenotype. The metabolome of an organism is complex and dynamic as the metabolites are getting continuously absorbed and degraded. Metabolomics studies attempt to provide a comprehensive snapshot of the physiological state of an organism at a given time state. Broadly, metabolomics study can be performed using two approaches: targeted and untargeted approach. In the case of untargeted approach, a number of different metabolites are measured without any sample bias, whereas, in the case of targeted approach, defined sets of metabolites are measured with an objective of the problem to be addressed. However, the steps in both these approaches are common. The first step is to outline the study design where the number of factors is taken into consideration like the sample size, randomisation, etc. This step is done to ensure that all important factors are considered addressing the metabolites involved and their putative interactions. The second step is the preparation of the sample, where the collection, storage and preparation of the sample take place. In the third step, an analytical technique like mass spectroscopy or NMR is utilised to measure and quantify the metabolites. The fourth step is to preprocess the data for analysis in order to extract biological inferences. This step is crucial to avoid noise in the data and perform background correction. The final step would be data analysis. This step includes applying statistical inferences to the data and clustering the data. The aim of this step is to perform the categorisation of the sample properties. Once a metabolomics study is completed, it can be subjected to various applications since it is an approach that is most proximal to capture the phenotype of an individual. This makes it an invaluable tool for pharmaceutics and healthcare. Advanced areas like personalised medicine utilise the metabolomics study for medical diagno-

A. Sengupta (✉) · P. Narad
Amity University, Noida, Uttar Pradesh, India
e-mail: asengupta@amity.edu; pnarad@amity.edu

© Springer Nature Singapore Pte Ltd. 2018
P. Arivaradarajan, G. Misra (eds.), *Omics Approaches, Technologies And Applications*, https://doi.org/10.1007/978-981-13-2925-8_5

sis and prognosis useful for the identification of the disease. Metabolomics is useful as it has the capability of identification and characterisation of different metabolites, making us understand the disease mechanisms in a better way.

**Keywords** Metabolomics · Flux analysis · Metabolic networks · Metabolic interaction

## 5.1   Metabolites to Metabolome

Metabolism is regulated by biochemical regulators such as metabolites. The central dogma of biology information flow from the gene to the transcript, to the protein and to the metabolite. An example of how metabolites regulate metabolism is glucose regulation in humans. Therefore, we need to study metabolism by applying the scientific technique called metabolomics. This field deals with the quantitative and qualitative study of the low molecular weight chemicals involved in metabolism called metabolites (e.g. carbohydrates, amino acids, fatty acids and hormones). These metabolites may be present within cells or in the environment surrounding cells and tissues, such as in blood or in urine or even whole organisms, and provide a better understanding of cellular biology. In a nutshell, we can define the field as the comprehension and understanding of the small molecules such as metabolites and other bio-fluids in a given tissue or organism known as metabolomics. Conjointly, metabolome is the interaction between these small molecules within a biological system. The metabolome is the complete collection of metabolites that helps to establish a link between the genotype and phenotype.

Metabolomics is an interdisciplinary field, and it requires the input and data information from all other fields of 'omics' such as through genomics, through transcriptomics and partly through proteomics. Its focus is on the study of the metabolic components such as enzymes, substrates and products. As a biological system, these 'omics'-based approaches are generally influenced by intrinsic factors such as genetic and extrinsic factors such as the environment factors. The term metabolomics is derived from two words: metabolites mean small biological molecules and omics means measures. It directly reflects the elementary biochemical activity and state of cells or tissues. It uses recent sophisticated analytical techniques to identify and quantify cellular metabolites with the application of statistical methods for data interpretation. Global analysis of substantial number of cellular metabolites can be done through metabolomics. It generates copious amounts of data using sophisticated statistical, mathematical and bioinformatics tools. Metabolomics has many applications in the study of gene ontology, data analysis, raw analytical data processing, information management and systems biology. Metabolomics is quite useful for analysing the overall effects of transcriptional and genetic manipulations, comparing mutants, analysing growth curves and assessing responses to environmental stress, toxicology, nutrition, study of cancers and diabetes, drug discovery and natural product discovery.

Previous literature has advocated the role of metabolomics in the field of in silico biology and drug discovery pipelines. The advent of the next-generation sequencing era has led to a better understanding of disease mechanisms and lays the foundation of personalised medicine. Metabolomics has made the metabolites an ideal biomarker by making them visible in a majority of biological and medical databases. It is also beneficial in getting a snapshot of the biological system as a whole. It is important to study metabolomics to combine and comprehend the data being generated from other 'omics' sciences (Riekeberg E et al. 2017).

Metabolomics provides a direct and sensitive measure of the phenotype at the molecular level. It provides an amplified and dynamic measure of changes resulting from processes involving the genome, transcriptome, proteome and the environment.

We need to measure the combined effect of the genome, lifestyle and our environment, on how we function which provides us the analysis of the phenotype by both hypothesis generation and hypothesis testing. Hence metabolomics provides a direct and sensitive measure of the phenotype at the molecular level. It provides an amplified and dynamic measure of changes resulting from processes involving the genome, transcriptome, proteome and the environment.

Besides these, metabolomics is also useful in pharmacology, plant biotechnology, crop breeding and toxicology. To understand the networks underlying complex biological processes, metabolomics can be perceived as a defined methodology for the study of functional genomics. It uses techniques of mass spectroscopy and nuclear magnetic resonance spectroscopy to identify transcriptome and proteome at systems level. Since metabolites are dynamic in nature in terms of time and space, they require special analytical procedures in their measurement. The state of metabolites within a biological system, gene expression and regulation, regulation of enzymes and their role in metabolic reactions can also be determined with the help of metabolomics. It reflects changes in functioning of a tissue or organism and their phenotypic expression as compared to genomics and proteomics. The techniques involved in metabolomics are relatively inexpensive, automated and rapid as compared to genomics and proteomics techniques. It is also useful in pathological study of human diseases such as diabetes, cancer and coronary and autoimmune diseases. Metabolomics has been used in many fields of application as it can provide beneficial tools such as food technology, microbial biotechnology, toxicology, enzyme discovery, plant biotechnology and systems biology. Various natural products generating from plant sources which include pharmaceutical sciences, food biotechnology and others that can be used in plant breeding and nutrition assessment have been developed with the extensive use of metabolomics. Metabolite profiling, metabolic fingerprinting and footprinting are some of the advanced applications of metabolomics.

## 5.2   Data Resources for Metabolomics

Metabolomics is redefining a new era of 'omics' sciences where the prime focus is shifting to the next-generation sequencing technologies for the recognition and characterisation of elements like small molecules/metabolites in the complete metabolomics content of an organism (Wishart DS. 2007). The term 'metabolome' refers to individual components of a metabolic system which are (<1500 Da) found in an organism. Metabolome can be treated as a counterpart to the terms 'genome', 'proteome' and 'transcriptome'. Metabolomics not just fills in as a foundation to frameworks science; it is starting to fill in as a foundation to different fields too. Metabolomics is beginning to hold importance in other areas of research such as systems biology and drug discovery.

Unlike its more developed 'omics' accomplices, metabolomics is as yet advancing a portion of its fundamental computational framework. The state of genomics, proteomics and transcriptomics is advanced, and the data is available online in the form of databases and repositories. However, metabolomics data resources are still in its infancy stage, and most of the data is still accessible through research papers and online journals only. Metabolomics additionally contrasts from other 'omics' sciences as a result of its solid accentuation on chemicals and systematic science procedures like NMR, mass spectrometry and chromatography. Thus, any online tool or database available for metabolomics is not very advanced, and there is a lot of scope of development of new software and platforms to analyse the metabolomics data. The resources available should not only be dealing with the small molecules and the metabolites but also should be able to represent the complete picture of the metabolome. Thus it implies that more resources should be developed which integrate information from other 'omics' based approaches and combine it with the metabolomics data.

However, within the past decade, we have seen a number of software programmes/resources that have been developed as a part of computational metabolomics. These resources are based on the need to integrate the data being generated in different laboratories and make it publicly accessible to the scientists throughout the world. The developing consensus is the requirement of required laboratory information management systems (LIMS), which will help the researchers manage their own data and access the data being generated through different laboratories for a better understanding of the process. The last decade has seen improvements in different areas of computational metabolomics such as (i) development of metabolomics databases, (ii) development of data standards and LIMS, (iii) development of tools useful for spectral data analysis and (iv) platforms for the metabolic modelling and visualisation.

Since this field deals with identifying a few metabolites at any given time and further utilising these metabolites in the process of marker identification for diseases, researchers working on metabolomics require online resources that can be accessed by metabolic pathways, names of the compound, X-ray, NMR spectra and mass spectrometry and also chemical structures and properties. Also, scientists need to look for metabolite properties, tissue or organ locations and metabolite association with diseases. The databases consist of information on different cadres of metabolomics. The primary information that can be extracted is about the compounds in the reaction and the diagrammatic representation of the reactions. Next, the online databases are freely available which consist of the concentration of the metabolites and the location of substrates in the organism. Further, information can be extracted on the subcellular locations and the physiochemical properties. Few of the databases also provide us information on the nomenclature of the enzyme through enzyme classification (EC) numbers. The chief characteristics of these data resources are that they are easily accessible and the information contained is validated through experimental procedures. These data resources are also completely referenced providing a link to the literature cited through them. Some of these also contain integrated platforms for the analysis of the data and enable easy interpretation of the metabolomics data.

## 5.2.1  EMBL-EBI

The efficient investigations of micromolecules and metabolites in a tissue, cell, cell culture or bio-fluid that are substantial after-effects of cell procedures of an abiotic stress that are recorded in the databases. The unique bit of knowledge into metabolic procedures that are occurring in cell condition are accumulated by identification and characterisation of such metabolites. Further, metabolic profiles that are taken from organic liquids can possibly go about as biomarkers for various sicknesses, for instance, diabetes and heart disorders, and impacts of diet routine (Whitfield PD et al. 2004). Extension of knowledge in natural research zones, for example, metabolic displaying and frameworks science, pharmaceutical research, toxicology and sustenance, is yielded from metabolomics advancements. In any case, analysts require access to learning and information to influence derivations and contrast from the outcomes the results they obtain in experiments to harness full potential of metabolomics (Haug K et al. 2012). The metabolome is the general supplement of metabolites under given dietary, hereditary or natural conditions which show in organic conditions. As of late, a few instruments or animal varieties particularly metabolic databases have been made to gather various trials together for an offered animal types to precisely mirror the basic decent variety and multifaceted nature. The major resources for these are Human Metabolome Database [HMDB] and Biological Magnetic Resonance Data Bank [BMRB]. At EMBL-EBI, the name of the resource which provides access to the data from all over the world in a single framework is called as MetaboLights. This database helps in facilitating the use of a common platform for sharing the different formats and also helps in the reproducibility of the data across the researchers all over the world (Salek RM et al. 2013). Currently, MetaboLights is composed of two layers of information: the first layer is the repository, and the second layer is the reference layer. This database is a part of the major standardised platforms such as the Metabolomics Society and the Metabolomics Standards Initiative (MSI) (Bino RJ et al. 2004). It is expected that in

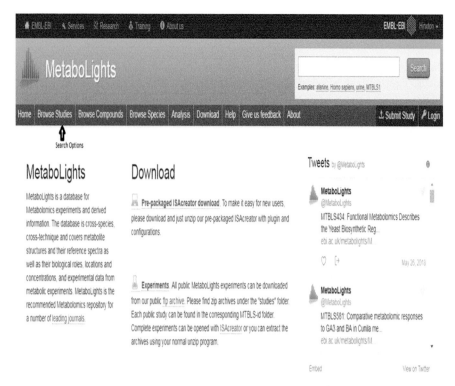

Fig. 5.1 Homepage of MetaboLights at EMBL-EBI web server

the future this database will be collaborated with the Reactome database in order to provide a more comprehensive view of the reactions (Fig. 5.1).

## 5.2.2 BRENDA

BRENDA (the Comprehensive Enzyme Information System) is one of the most widely used and comprehensively built resources for the enzymes. It consists of information on the enzymes based on the molecular and biochemical experiments conducted on the enzyme.

Enzymes are the most diverse category of proteins and they are the largest known proteins in the database. Enzymes catalyse most of the chemical reactions in the organism related to metabolism. The enzymes have a crucial role in the regulation of metabolic reactions within the organism. The enzyme data has increased many folds with the invention of new techniques for structural and functional genomics. The need of the hour is to understand the enzyme data and use it in an effective manner to enhance our understanding of the metabolic networks. One such resource, which helps us with the information on enzymes and other metabolic data, is

BRENDA. The database consists of information which is extracted from the literature (Schomburg I et al. 2002).

BRENDA can be considered as a comprehensive resource which comprises the classification of the enzymes on the basis of the EC (enzyme classification) numbers. The classification recorded is based on the reaction type, i.e. whether it is an oxidation or reduction reaction or whether it is classified as hydrolysis or the group transfer. BRENDA database is based upon not just a single organism of specific set of enzymes; rather it covers the organism-specific information about the enzyme names, the type of reaction catalysed, the kinetics behind the reaction, the number of substrates and products involved and the molecular and functional properties of the enzymes.

In its current version, BRENDA consists of records of 4200 EC numbers that are related to 83,000 different enzymes. The records in the database are updated regularly through the literature curation from the research papers from resources of repute such as PubMed and SciFinder.

One of the other most important aspects of BRENDA is that it also contains information on the ligands. These are those compounds that have a function in modulating the activity of the substrates or the products or the cofactors. Currently, BRENDA has a record of ~3,20,000 enzyme-ligand relationship. A total of 33,000 different compounds are present in BRENDA which is listed as ligands.

BRENDA is also useful as it allows the user to perform simulation of the metabolic networks. This can be done through BRENDA by incorporating the kinetic data and the enzyme information from other resources such as KEGG. The primary assumption that has to be made for performing simulation is that the metabolic networks must be treated as directed graphs. BRENDA has also included the information on the human diseases as the literature is increasing many folds. Literature curation was done through PUBMED by looking for keywords related to that particular enzyme (Fig. 5.2).

## 5.2.3   HMDB

The Human Metabolome Database (HMDB) is one of the most widely used resources for providing comprehensive information about the small molecule like the metabolites for the human system. It provides high-quality data which is freely accessible. It was created under the aegis of the Human Metabolome Project which was an initiative of the Genome Canada (Xia J et al. 2009). It can be considered as one of the first and dedicated resources for metabolomics data. HMDB is able to help the scientist in metabolomics research by providing data on the identification of metabolites using NMR, LC/MS and other mass spectroscopy-based approaches. The architecture of HMDB can be divided into three sections: (i) chemical data, (ii) molecular data and (iii) clinical data (Danaher J et al. 2016).

The latest release of HMDB is the HMDB 4.0 which has advanced in number of ways since the inception of HMDB 1.0, which was first released in 2007. The first version was released with data from 2180 human metabolite on the basis of the physiochemical and biological data. In 2009, the second version was released as

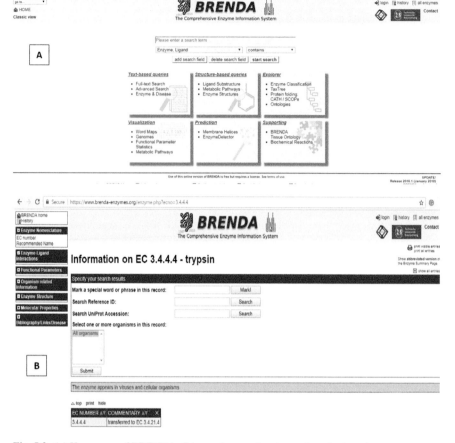

Fig. 5.2 (a) Homepage of BRENDA, (b) search page showing options for trypsin

HMDB 2.0, which added more annotation most of which came from the NMR data. Literature annotation was further added and the second version consists of information from 6408 human metabolites. In 2013, the third version was released as HMDB 3.0. This version was more advanced from the others as it also incorporated data from the lipid and food information. This version also expanded the information content in the HMDB's spectral library and incorporated data from the metabolic pathways. In this version, the interface also was improved substantially, and now the content was expanded to 40,153 metabolites.

The latest release was introduced in 2018 by the name of HMDB 4.0. This is by far the most advanced and updated release of the consortium. This version has been produced by the using bioinformatics approaches and literature curation. The final version now has 1,14,100 human metabolite information which is divided into three categories: known metabolites, predicted metabolites and expected metabolites. Another salient advancement is the increase in the number of reference spectra from NMR and mass spectroscopy approaches. This data has also now reached to 3,51,754 which includes information from both computational and experimental observations. This increase can be considered to be 100-fold than what was present in the previous versions of the database. The next advancement was seen in the pictorial representation of the metabolic pathways. The representation has undergone a 60-fold increase as the total number of representations has now increased to 25,770. The metabolite and disease associations have also witnessed increase of 77% to a total number of 5498. The most significant improvement which is important for the researchers is the increase in the number of metabolites. As we discussed in our previous paragraph, in the first version in 2013, the metabolites were classified into three groups: (i) detected metabolites which have been quantified, (ii) detected metabolites which have not been quantified and (iii) predicted or expected metabolites. The first category consists of the small molecules that have been identified and for whom the concentrations have been quantified, and also through experimental records, they have been proved to be present in the human system (Allen F et al. 2014). Those small molecules which fall under the 'expected' category are those for which the structure elucidation has been done, but there is still no experimental evidence of the presence of such metabolites in the human system. As a result, the maximum number of metabolites in the latest version is under the category of 'expected' as being 82,274 in HMDB 4.0. As a consequence, the number of metabolic pathway representations has increased many folds in HMDB 4.0. The latest version also consists of a new class of compounds known as the 'predicted' compounds. This category was introduced because of the scarcity of the metabolomics data pertaining to identification of the metabolites. These compounds can be easily filtered by the users as per their requirements. As a result of the number of entries increasing in the HMDB 4.0 version, the accession number has now accentuated to a five-digit accession number format. This version also consists of ontological definitions for the metabolites. These ontological definitions have been fully standardised and consist of more comprehensive hierarchical annotations. The ontology definitions are represented under four categories: (i) process in which they are involved, (ii) function of these metabolites, (iii) the physiological effect of these metabolites and (iv) the disposition data. Another striking feature of HMDB 4.0 is the inclusion of pharmaco-metabolomic data and single nucleotide polymorphism data.

HMDB 4.0 also accomplishes enhancements in terms of the user interface. The visualisation of the metabolic pathways has improved significantly with better viewing options, improved searching tools and a number of options for the exchange of data across different platforms (Law V et al. 2013). Another important feature is that all the outputs have hyperlinks to important repositories such as UniProt and DrugBank which makes it easier to link the information across different resources. The output consists of a summary of the results which helps the user to infer the observations in a more effective manner. The chemical structures can be easily retrieved in a number of known formats such as SMILES, SDF, PDB, etc. The other information such as the sequence information of DNA/proteins can easily be accessed through the standard FASTA format. All of these HMDB 'component' files can be freely downloaded from the HMDB 4.0 website.

**Fig. 5.3** (a) Homepage of HMDB, (b) search page, (c) search page of metabolites

In a nutshell, since its inception from 2013, HMDB has increasingly become a comprehensive specialised database for metabolomics all over the world. In particular, the latest version HMDB 4.0 has undergone a number of improvements and is useful for 'predicted' metabolites data and the prediction retention time information (Fig. 5.3).

## 5.2.4   SABIO-RK

The understanding of a biological system needs the systemic study of the individual components as well as the overall comprehension of the interactions between individual components. Metabolic modelling is one such field which can help us in getting a complete picture of the biological process in the form a network diagram. For the successful analysis underlying metabolic networks, it is important that we have correct kinetic data for each reaction so that we can model the enzyme kinetics. SABIO-RK is an important resource to accomplish this goal in metabolic modelling.

SABIO-RK was introduced to the scientific community as a database which stored information on kinetics of metabolic reactions. This information is useful for the scientists performing metabolic modelling, and also it helps the experimental researchers to comprehend complex metabolic pathways easily in a network representation (Wittig U et al. 2011). In comparison with other data resources, SABIO-RK consists of reaction-based visualisation for the dynamics of the reaction. Also, the database consists of elaborate information of the pathways regarding the location in the cell, the constituents of the reaction and also the complete information about the enzymes involved in the reaction.

SABIO-RK consists of an integrative pipeline where the data can be consolidated from the literature and also from the experimental observations (Dräger A et al. 2014). Thus, this database shows a better output and is highly flexible in its utility because of the advanced interface and better options for the user. The search engine has the option of querying the organism based on the information from NCBI, BRENDA and ChEBI, thus offering more flexibility and better usability.

SABIO-RK further also helps in integration of data from various sources which help in establishing a broad-based information resource (Funahashi A et al. 2007). The complete information is extracted from literature under the headings of reactions, substrates, products, details of the catalysts, the conditions in which the reaction takes place and the details of the paper from which this information is extracted. Further, the information is enriched through storing information about the enzymes that catalyse the reaction. Detailed information is also available for the isozymes or mutations that were used in the experiments, also about the subunit compositions and accession numbers from the UniProt database. The database also holds information on the kinetic data for every elementary step and a visual representation of the mechanism of the reaction.

In totality, SABIO-RK has kinetic information for more than 650 species constituting 7250 metabolic reactions and a repository of approximately 1000 enzymes. Another striking feature is the link of the reactions (~2400) to the KEGG Ligand database. There are also options to submit kinetic data to SABIO-RK.

Kinetic data can be inserted into SABIO-RK in two different ways. The data that is generated through the literature can be incorporated by the researchers all over the world by using the online interface of SABIO-RK (Krebs O et al. 2007). The data that is generated from the experiments can be submitted through their submission platform and the format used is SabioXML. Hence, the process of submission is automated, and it helps in making the process enhanced.

The online resource of SABIO-RK is protected by a password and thus is useful for the researchers to store the data into a local database. The web interface consists of a variety of options ranging from form fields and structured input data. The consistency is maintained by the automated process of generation of metabolic reactions and their equations, which cannot be manually changed.

In order to avoid discrepancies in terms of the reactions, substrates, locations, lists of parameters, organisms and types of kinetic laws, the existing information present in the SABIO-RK database is provided to the users (Wittig U et al. 2014). The controlled vocabulary comes from trusted sources such as the NCBI database, BRENDA and the ontology databases. The utility of controlled vocabulary is that along with the annotations, these help to identify the biological context of the metabolic reactions. Ontologies used for adding the biological context are retrieved from NCBI taxonomy, ChEBI and Systems Biology Ontology.

The SABIO-RK data can be accessed using the online servers and services that are freely available. These online resources support the XML/SBML standard format. SBML is a standard format for data exchange and can be easily used for the exchange of data amongst the different web servers.

The online interface is useful for the researchers to look for the metabolic reactions and apply kinetic laws to the equations. The interface is also useful for creating complex queries by using different identifiers such as the UniProt IDs, tissue names, cellular spaces enzymes, substrates or the pathways that are involved; a number of searches are based on the ontologies about the biological context, which are helpful in defining the relationship between the controlled vocabularies and the objects.

SABIO-RK has two different web interfaces currently. The previous SQL-based search was now changed to inverted indexing. The kinetic data records that are generated can now be seen on the screen to the users and simultaneously the queries can be formulated. This feature is useful for the users and enhances the capability of the web interface. An additional feature that has been added is the visualisation of the reaction-based information.

In summary, SABIO-RK is a data resource which consists of the metabolic reactions and their kinetic information. It is beneficial for the users as it helps them to model the metabolic reactions by adding the kinetic data and other details like the cellular locations, etc. The information on the kinetics of the reaction can be retrieved either from the literature by manual curation or the information can be

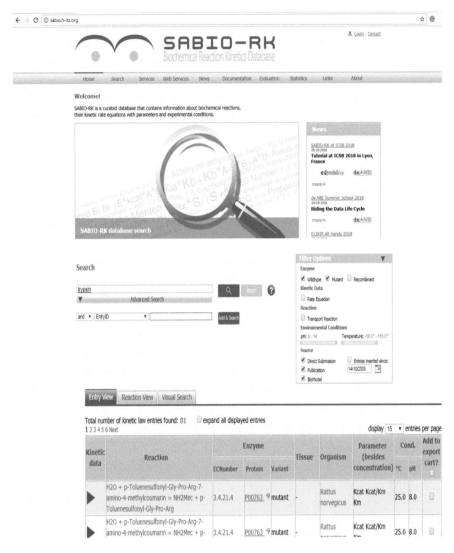

**Fig. 5.4** Homepage of SABIO-RK

added from the experimental approaches. The addition of annotation to controlled vocabularies and the information on biological ontologies enables the user to perform complex querying to the database. The datasets can be retrieved and exported to the SBML format from the SABIO-RK resource (Fig. 5.4).

The other resources are beyond the scope of this chapter and hence collated in the form of a tabular representation below (Table 5.1):

**Table 5.1** Various metabolomics data resources

| S.NO | Database Name | URL | Description/comment |
|------|---------------|-----|---------------------|
| 1. | BIGG | http://bigg.ucsd.edu/home.pl | Database of human, yeast and bacterial metabolites |
| | | | Database for pathways and reactions as well metabolic modelling |
| 2. | Biological Magnetic Resonance Bank (BMRB Metabolomics) | http://www.bmrb.wisc.edu/metabolomics/ | Emphasis on NMR data, no biological or biochemical data |
| | | | Specific to plants (*Arabidopsis*) |
| 3. | METLIN Metabolite Database | http://metlin.scripps.edu/ | Human specific |
| | | | Mixes drugs, drug metabolites together |
| 4. | Golm Metabolome Database | http://csbdb.mpimpgolm.mpg.de/csbdb/gmd/gmd.html | Emphasis on MS or GC-MS data only |
| | | | No biological data |
| | | | Few data fields |
| 5. | Fiehn Metabolome Database | http://fiehnlab.ucdavis.edu/compounds/ | Tabular list of ID'd metabolites with images |
| 6. | NIST Spectral Database | http://webbook.nist.gov/chemistry/ | Spectral database only (NMR, MS, IR) |
| | | | No biological data |
| | | | Little chemical data |
| | | | Not limited to metabolites |

## 5.3 Computational Approaches for Metabolomics Analysis

It is extremely challenging to understand different mechanisms of a living organism. The simplest way of understanding the functionality of the organism is to decipher the metabolism, which is an essential process to produce energy for the working of the cell. For a better understanding of the process, it is important to first comprehend how the energy is consumed and produced by the metabolites and how this process is regulated in an efficient manner. The entire process encompasses the use of enzymes that catalyse the reaction at each step.

The understanding of the processes involves the answering of the following questions:

(i) How and what are the inputs that the organism needs for the process?
(ii) What is the classification of different types of outputs?
(iii) What is the efficiency of this process?

The easiest way to deal with these questions is by the identification of the metabolic reactions involved in the processes. After we decipher the reactions involved in the organism, we can easily establish the operation of the metabolites. With the advent of the Human Genome Project, large number of data resources is available. In order to make sense of these data, it is endeavoured that in silico systems biology plays a major role. One such approach is modelling and reconstruction of the metabolic pathways in the form of a network-based representation. One of the most common of such approaches is the modelling of metabolic pathways using the graph theory and path finding using 'metabolic path finding'. The task at hand is to decide which path is to be followed in order to discard the false positives from the dataset and make the validation of the data possible. However, it is important to decide the best modelling approach when representing a metabolic network in the form of a biological system. The data under consideration should be understood in a comprehensive manner in order to provide a meaningful picture of the process. The second important criterion should be that the kinetics and other parameters that are added in the network should be correct and complete. This will enable us to make a proper representation of the metabolic system.

A general definition of a metabolic model is the one which is composed of metabolites and the reactions between the metabolites. A metabolic reaction can be unidirectional (i.e. going in one direction only) or bidirectional (i.e. going in both directions). When the reaction is defined as being bidirectional, the reactions occur at the same time, and thus equilibrium is defined as the point where the rate of conversion from the substrate to product and vice versa is the same. The ratio of the concentration can be quantified by using the equilibrium constant K; this can be known to be constant at a given temperature. The analysis of the metabolism as a complete process of networks, the reaction equilibrium has to be taken into consideration. However, some reactions never reach to an equilibrium state, and thus they are unidirectional in nature. These kinds of reactions are called as irreversible reactions. The metabolic compounds can be defined as nodes, and the reactions can be defined as the links/edges or the reverse known as the reaction graph. A bipartite graph is the one which uses the compounds and the reactions as nodes. The network can be classified further as weighted or unweighted and also as directed or undirected graphs (Fig. 5.5).

### 5.3.1 Network Analysis and Metabolic Pathway Integration

In the last decade, the field of biochemistry has enhanced with the amount of data increasing on chemical kinetics. With this increase in the amount of kinetic data, it has now become easier to understand the biochemical processes in the form of a network-based approach.

One such approach which has received attention in the last few years is the computational modelling of the metabolic pathways. In the following text, the stepwise protocol of the construction of the metabolic pathways is discussed. If we have a set of reactants which are parts of the metabolic pathway under consideration, atomic

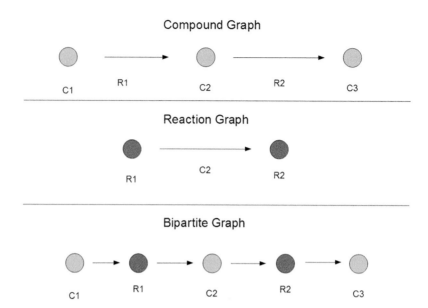

**Fig. 5.5** Representation of metabolic network as nodes and edges. The metabolic compounds can be defined as nodes, and the reactions can be defined as the links/edges or the reverse known as the reaction graph. A bipartite graph is the one which uses the compounds and the reactions as nodes

**Table 5.2** A summary of other metabolomics-based approaches is provided in tabular form

| Approach | Advantage | Disadvantage |
|---|---|---|
| Targeted analysis | Quantitative | Does not detect compounds that were not targeted |
| | High throughput | Limited number of compounds can be targeted |
| | Low limit of detection | Targeted compounds must be available purified for calibration |
| Metabolite profiling | Global (not targeted) | Semi-quantitative |
| | | Majority of peaks are not identifiable |
| | | Medium throughput |
| | | Difficult informatics |
| Metabolite fingerprinting | Global (not targeted) | No compound identification |
| | Directly applicable to pattern recognition | |
| | Highest throughput | |

components could be generated, and then these can be preserved for model composition. These compounds then can be used to compose the models, which can be evaluated through experimental observations of the reactants. Performing simulations can help in analysing the relationship between the substrates and the products for a given set of reactions (Table 5.2).

### 5.3.1.1  Metabolic Control Analysis

Metabolic control analysis is one of the most important approaches which is useful to determine the effect of the individual reactions of different biochemical pathways. This process makes the analysis in a quantitative manner. The metabolic control analysis is defined by two coefficients: first are the control coefficients, which help to characterise the system flow response and values of perturbations; second are the elasticity coefficients, which are useful in quantifying the rate of change of reaction after the simulations of kinetic values under specific conditions. This approach has been used for studying a number of genetic pathways. The differential expression of the enzymes can be studied using the metabolic control analysis. Metabolic control analysis has been applied to a number of researches in the past to perform kinetic modelling (De Matos P et al. 2009).

### 5.3.1.2  Stoichiometric Analysis

Stoichiometry is done on the law of mass conservation, which states that mass total of the reactants and the products should be equal. This leads to knowledge that the quantity of reactants and the product results in a ratio of positive number. Thus, it infers that if individual quantities of reactants are known, then we can calculate the quantity of the products.

## 5.3.2  Flux Analysis

Metabolomics refers to an analysis platform that involves distinguishing and evaluating all the small molecules and the metabolites in a given sample at a given time. These metabolites can be characterised using a non-targeted and unbiased approach which uses different kinds of systems such as the mass spectroscopy and NMR. These techniques can be combined with other separation techniques such as chromatography. Further statistical techniques are used for the analysis of the spectra and then compared with the known samples. The need of an additional approach came because some of the metabolites occur at low concentrations and hence are difficult to characterise.

Flux balance analysis (FBA) is one such method that has been presented as a displaying and examination apparatus for metabolomics (Orth JD et al. 2010). It is a limitation-based approach for demonstrating metabolic systems. FBA uses straight enhancement to depict the enduring condition of the response motion conveyance in a metabolic framework by characterising a goal work, for example, the development rate or the generation of ATP. The relentless state examination is done by FBA by making utilisation of the stoichiometric lattice for the framework (Bordbar A et al. 2014).

### 5.3.2.1  Incorporation of Additional Constraints

Different methodologies adjusting extra limitations have been prescribed to represent quality direction and free vitality. The statement of every quality is directed by the quality articulation. The effect of transcriptional direction on cell digestion was dictated by the Boolean coherent administrators. The administrative tenets for a given quality are related with an arrangement of Boolean articulations. In the event that the principles are satisfied, the quality is 'on', and the relating protein and its related response partake in the system. The Boolean principles are gotten from trial work describing administrative procedures, e.g. microarray information. Joining of response thermodynamics in FBA done by the burden of extra non-straight requirements delineating vitality adjusts with the concoction potential (Covert MW et al. 2001). The vitality adjust plan was extended further to maintain a strategic distance from thermodynamically infeasible cycles from a given response arrange. In this approach, the duplication amongst transitions and related compound potential prompts computational flightiness for significant scale frameworks.

### 5.3.2.2  Interfacing Metabolomics with FBA

FBA has contributed a noteworthy part in the investigation of natural systems and to make expectations utilising testable analyses with cell practices. One of the significant difficulties of metabolomics is that it needs thoroughness and exactness in estimations. Regardless of these difficulties, there has been a considerable measure of progressing advancements of various detachment and scientific strategies, affectability and selectivity of estimations turn into key worries for substantial system frameworks where concurrent examination of several metabolites is required (Beard DA et al. 2004). The metabolomics information in relationship with FBA tries to fathom a portion of these difficulties.

The impediments of metabolomics can sufficiently be supplemented by a FBA-based formalism. The joining of fractional metabolic data has been permitted by the use of genome-scale reproductions. Estimated transitions or motion proportions can be forced as equity imperatives. Moreover, the dynamic profile of focuses can be used helpfully for estimation of motion data, i.e. $v = dC/dt$. The vulnerabilities that are connected with fixations can be assessed by mirroring their conviction limits (Fischer E and Sauer 2003) in imbalance imperatives. FBA alongside a few numerical examinations, for example, affectability investigation [68] and target work induction instrument (Burgard AP et al. 2003), can coordinate tests facilitated by perceiving critical metabolites. A particular diagnostic technique required by the undetected or vital metabolites would thus be able to be distinguished and bottlenecks in the consistent systems can be alleviated effectively.

## 5.4   Applications

The applications of metabolomics are highly diverse, and thus it is touted to be as one of the most important tools for the comprehension of the living organisms and the study of diseases and disease mechanisms. The applications are becoming more diverse in the times to come, and hence the field is providing important insights to the researchers as well as clinicians all over the world. This field of metabolomics is now widely used in the research related to disease identification, drug discovery and development, nutrigenomics and agricultural research (Fig. 5.6).

### 5.4.1   Toxicology

Metabolomics applications can be well studied for the field of assessment of toxicity. The metabolic profiling of the urine samples or the blood samples can be used for the determination of the toxicity levels. A number of different techniques are available for the study of metabolic profiling as discussed above. This approach can also be used for analysing the disease conditions associated with the liver and the kidney. Recently, pharmaceutics have also advocated the metabolomics technologies since they find it the most useful approach to detect the toxicity levels

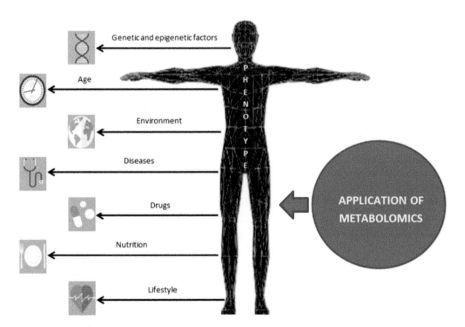

**Fig. 5.6** Applications of metabolomics. It is now widely used in the research related to disease identification, drug discovery and development, nutrigenomics and agricultural research

associated with the drug candidates. Thus, it would be beneficial in saving tremendous amount of resources and funds.

## 5.4.2 Functional Genomics

Metabolomics can also be of great utility in the research of the phenotypes which are resulting from the genetic changes through the field of functional genomics. To elaborate, it can be suggested that the knowledge about the metabolome would be useful to predict whether a gene was affected by insertion/deletion in the genome of an organism. The identification of the phenotypic effects can also be utilised to a number of other applications. For instance, genetically modified organisms such as GM crops can be detected for phenotypic changes after human consumptions. These changes have the potential in principle to change the metabolome information of the organism, and hence a prior prediction of the phenotypic changes associated with their consumption would be useful to avoid any harm from such crops.

Metabolomics can also aid in the prediction of the function of the unknown genes. This is possible through the comparison of known metabolic profiles with that of unknown genes and transferring the orthologous information. Model organisms of *Saccharomyces cerevisiae* and *Arabidopsis thaliana* are currently being undertaken and could lead to such advances in the future.

## 5.4.3 Nutrigenomics

With all the 'omics' knowledge in place such as the transcriptomics, genomics, proteomics and metabolomics, these knowledge can be transferred to the nutritional principles in human, and this field is known as nutrigenomics. There are two types of factors that affect the metabolites: the first category is of the endogenous factors such as the gender, body composition, genetic conditions and age; the second category is of the exogenous factors which include the food composition, the medications taken and the nutrients required. The field of metabolomics can be applied to nutrigenomics in determining the metabolic fingerprint of the organism, which will provide the complete snapshot of both the factors and their effect on the individual system.

The combination of knowledge of various field of biology such as transcriptomics, genomics, metabolomics and proteomics using analytical techniques with nutritional principles for humans is called nutrigenomics. Factors affecting the metabolite are divided into two types.

### 5.4.4 Health and Medicine

Metabolomics also has an important role in the understanding of the disease mechanisms and the treatment of a number of different health conditions affected by the metabolic changes in the genome such as cancer. This field can be used to identify the pathophysiological states of the disease and help in understanding targeted disease mechanism. For instance, the identification of metabolomics biomarkers can help us categorise the progression of several types of cancers.

### 5.4.5 Environment

Metabolomics can also be applied to characterise the ways in which an organism interacts with its environment. Studying these environmental interactions and assessing the function and health of an organism at a molecule level can reveal useful information about the effect of environment on an organism's health. This can also be applied to a wider population to provide data for other fields of research, such as ecology.

### 5.4.6 Agriculture

Metabolomics helps us to improve the genetically modified crops and also helps us to identify the dangers of the consumption of GM crops by making us get a snapshot of the plant development at different time frames. Plant metabolite identification is particularly important as it would help us in identification of the functions of the primary and secondary metabolites.

### 5.4.7 Biomarker Discovery

Another field where metabolomics plays an important role is the biomarker identification through decision making. Biomarker identification is an important area of research for disease diagnosis and understanding. Using metabolomics, the biomarkers can be considered as the metabolites, which can be used for the classification of two groups of samples: the disease group and the healthy group. Biological samples from the bile, urine or seminal fluids are important source of metabolic information, and this information can be processed though metabolomics by metabolic profiling or fingerprinting for the identification of biomarkers.

### 5.4.8 Personalised Medicine

The new era of medicine is going to be personalised medicine, which holds great promise in the field of healthcare. This field would aid in early diagnosis and treatment of diseases. Classical biochemical tests are used for disease identification; this would to go another level with incorporation of metabolomics approaches. Individual samples from the metabolites and their concentrations could be used for disease diagnosis and treatment. The response of the individual to certain medications can be detected by checking the metabolite concentrations.

# References

Allen, F., Pon, A., Wilson, M., Greiner, R., & Wishart, D. (2014, June 3). CFM-ID: A web server for annotation, spectrum prediction and metabolite identification from tandem mass spectra. *Nucleic Acids Research, 42*(W1), W94–W99.

Beard, D. A., Babson, E., Curtis, E., & Qian, H. (2004, June 7). Thermodynamic constraints for biochemical networks. *Journal of Theoretical Biology, 228*(3), 327–333.

Bino, R. J., Hall, R. D., Fiehn, O., Kopka, J., Saito, K., Draper, J., Nikolau, B. J., Mendes, P., Roessner-Tunali, U., Beale, M. H., & Trethewey, R. N. (2004, September 1). Potential of metabolomics as a functional genomics tool. *Trends in plant science, 9*(9), 418–425.

Bordbar, A., Monk, J. M., King, Z. A., & Palsson, B. O. (2014, February). Constraint-based models predict metabolic and associated cellular functions. *Nature Reviews Genetics, 15*(2), 107.

Burgard, A. P., & Maranas, C. D. (2003, June 20). Optimization-based framework for inferring and testing hypothesized metabolic objective functions. *Biotechnology and Bioengineering, 82*(6), 670–677.

Covert, M. W., Schilling, C. H., & Palsson, B. (2001, November 7). Regulation of gene expression in flux balance models of metabolism. *Journal of Theoretical Biology, 213*(1), 73–88.

Danaher, J., Gerber, T., Wellard, R. M., Stathis, C. G., & Cooke, M. B. (2016, January 1). The use of metabolomics to monitor simultaneous changes in metabolic variables following supramaximal low volume high intensity exercise. *Metabolomics, 12*(1), 7.

De Matos, P., Alcántara, R., Dekker, A., Ennis, M., Hastings, J., Haug, K., Spiteri, I., Turner, S., & Steinbeck, C. (2009, October 23). Chemical entities of biological interest: An update. *Nucleic Acids Research, 38*(suppl_1), D249–D254.

Dräger A., & Palsson, B. Ø. (2014, December 8). Improving collaboration by standardization efforts in systems biology. *Frontiers in Bioengineering and Biotechnology, 2*, 61.

Fischer, E., & Sauer, U. (2003, March 1). Metabolic flux profiling of Escherichia coli mutants in central carbon metabolism using GC-MS. *The FEBS Journal, 270*(5), 880–891.

Funahashi, A., Jouraku, A., Matsuoka, Y., & Kitano, H. (2007, January 1). Integration of CellDesigner and SABIO-RK. *In Silico Biology, 7*(2 Supplement), 81–90.

Haug, K., Salek, R. M., Conesa, P., Hastings, J., de Matos, P., Rijnbeek, M., Mahendraker, T., Williams, M., Neumann, S., Rocca-Serra, P., & Maguire, E. (2012, October 29). MetaboLights— An open-access general-purpose repository for metabolomics studies and associated metadata. *Nucleic Acids Research, 41*(D1), D781–D786.

Krebs, O., Golebiewski, M., Kania, R., Mir, S., Saric, J., Weidemann, A., Wittig, U., & Rojas, I. (2007, March 1). SABIO-RK: A data warehouse for biochemical reactions and their kinetics. *Journal of Integrative Bioinformatics, 4*(1), 22–30.

Law, V., Knox, C., Djoumbou, Y., Jewison, T., Guo, A. C., Liu, Y., Maciejewski, A., Arndt, D., Wilson, M., Neveu, V., & Tang, A. (2013, November 6). DrugBank 4.0: Shedding new light on drug metabolism. *Nucleic Acids Research, 42*(D1), D1091–D1097.

Orth, J. D., Thiele, I., & Palsson, B. Ø. (2010, March). What is flux balance analysis? *Nature Biotechnology, 28*(3), 245.

Riekeberg, E., & Powers, R. (2017). New frontiers in metabolomics: From measurement to insight. *F1000Research, 6*, 1148.

Salek, R. M., Haug, K., Conesa, P., Hastings, J., Williams, M., Mahendraker, T., Maguire, E., Gonzalez-Beltran, A. N., Rocca-Serra, P., Sansone, S. A., & Steinbeck, C. (2013, January 1). The MetaboLights repository: Curation challenges in metabolomics. *Database, 2013*, bat029.

Schomburg, I., Chang, A., & Schomburg, D. (2002, January 1). BRENDA, enzyme data and metabolic information. *Nucleic Acids Research, 30*(1), 47–49.

Whitfield, P. D., German, A. J., & Noble, P. J. (2004, October). Metabolomics: An emerging post-genomic tool for nutrition. *British Journal of Nutrition, 92*(4), 549–555.

Wishart, D. S. (2007, July 11). Current progress in computational metabolomics. *Briefings in Bioinformatics, 8*(5), 279–293.

Wittig, U., Kania, R., Golebiewski, M., Rey, M., Shi, L., Jong, L., Algaa, E., Weidemann, A., Sauer-Danzwith, H., Mir, S., & Krebs, O. (2011, November 17). SABIO-RK—Database for biochemical reaction kinetics. *Nucleic Acids Research, 40*(D1), D790–D796.

Wittig, U., Rey, M., Kania, R., Bittkowski, M., Shi, L., Golebiewski, M., Weidemann, A., Müller, W., Rojas, I. (2014, January 1). Challenges for an enzymatic reaction kinetics database. *The FEBS Journal, 281*(2), 572–582.

Xia, J., Psychogios, N., Young, N., & Wishart, D. S. (2009, May 8). MetaboAnalyst: A web server for metabolomic data analysis and interpretation. *Nucleic Acids Research, 37*(suppl_2), W652–W660.

# Chapter 6
# Microbiome

**Debarati Paul, Sangeeta Choudhury, and Sudeep Bose**

**Abstract** The advancement in current experimental science and its technology have made it possible to understand the interactions of microbes with host, biotic and abiotic factors, and host responses. It has been realized that a single microorganism on its own is insufficient to cause disease in human/plant/animals, unless it is supported by the surrounding environmental factors and the existing mini-ecosystem consisting of various other microbes that may play antagonistic or synergistic roles. Although, in present scenario, one cannot predict or specify the genera, classes, or species of microbes that regulate a disease phenotype, the conclusions drawn from several experimental and human studies strongly suggest the presence and/or the levels of specific microbes comprising a population that govern the host phenotype. The metabolic networking elucidates interplay of metabolomics and metagenomics revealing the correlations between host and gut bacteria in health and disease conditions. This chapter summarizes the dynamics of microbial associations and a mechanism of divergent actions connecting the microbiome prevalent in environmental conditions (soil/marine-to-plant) leading to diverse health concerns.

The shaping of host-immune responses as well as modulating effects caused by interaction with drugs is linked with alterations in composition and diversity of microbial community in several studies. However, many questions will remain unanswered before we can bring out the full prognostic and predictive potential utilization of microbiomes.

**Keywords** Microbiota · Gut · Diversity · Prebiotics · Probiotics · Microbial infection · Host interactions · Pathogenesis

---

Authors Debarati Paul and Sangeeta Choudhury have equally contributed to this chapter.

D. Paul · S. Bose (✉)
Amity University, Noida, Uttar Pradesh, India
e-mail: drdebaratipaul@gmail.com; sbose1@amity.edu

S. Choudhury
Research Department, Sir Ganga Ram Hospital, New Delhi, India
e-mail: dr.sangeeta.sgrh@gmail.com

© Springer Nature Singapore Pte Ltd. 2018
P. Arivaradarajan, G. Misra (eds.), *Omics Approaches, Technologies And Applications*, https://doi.org/10.1007/978-981-13-2925-8_6

## 6.1   Introduction

Over the last decade, there has been a huge shift in the popular perception of micro-organisms—instead of considering them as potentially pathogenic organisms that should be destroyed, we now realize that the microorganisms living in and on us are an essential part of us and necessary for good health. Medical literature has compared humans to ecosystems (rather, equated as live ecosystems) where eukaryotic cells of our human body and bacterial/viral prokaryotic cells living inside and outside our bodies are intricately interwoven, and, consequently, the importance of the microbiome and interactions with the host has been realized and accepted to play significant roles in human health (Blaser 2016). The metabolites produced by these microorganisms interact with the immune system, the neuroendocrine system, and the digestive system, thereby directly affecting our physical and mental well-being.

In the case of agriculture too, a lot of importance is now laid upon the study of the whole microbiome instead of individual microbes that favor (nitrogen fixers) or disrupt (pathogenic strains) plant growth and productivity, as it is realized that the dissemination and spread of these cultures are based on interactions with other members of the microbiota and on the biotic and abiotic environment. Like the human microbiota, plant, marine, and soil microbiota are very complex and varied in different layers/parts (Fig. 6.1) and continuously influence each other and get impacted by the abiotic factors or biotic and human interventions.

## 6.2   Microbe to Microbiome

Recent years have witnessed the shift of research interest on individual microbes to "the entire range of microorganisms dwelling in a particular niche/environment" be it soil, plant, human, or aquatic, especially the marine environment. It has been realized that the abundance and/or scarcity of each individual strain depends upon the influence of the other accompanying strains apart from biotic/abiotic/human interventions. Some of them are commensals (freely live together), some are symbiotic (depends on each other for metabolism), and others are antagonistic (repel or kill); together, they form a "mini-ecosystem" in every niche that they inhabit.

As infants or even before, the mother's microbiome trains the child microbiota, and it develops and changes until old age depending upon the lifestyle, food, drugs, and physical state. The ever-changing nature of the microbiome is not only applicable to humans but also true for soil, plants, and aquatic microbiomes. Here we will focus on a few types of microbiome that are important for humankind and have been exhaustively studied and reported.

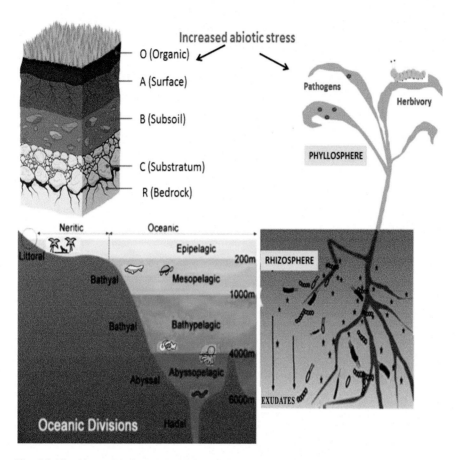

**Fig. 6.1** The biogeochemical cycles influenced by the soil microbiota. Soil, marine, and plant environments exhibit different layers or horizons that are impacted by various biotic and abiotic parameters contributing toward variations in microbiota

### 6.2.1 Soil Microbiome

The soil microbiome describes the microbes existing in soil, e.g., prokaryotes (bacteria, archaebacteria, and viruses) and micro-sized eukaryotes. The carbon and nitrogen content plays a huge role in defining soil microbiota of a particular niche. The microbial load or biomass carbon in soil may be >1000 kg per hectare of land (Serna-Chavez et al. 2013). Bacterial and fungal colonies dominate soil population contributing to 102–104 times more biomass than other occupants of the soil microbial species (protists, archaea, and viruses). The soil environment comprises vivid environments that exhibit differences in abiotic factors and microbial density and

**Table 6.1** Factors that interact and help in the structuring of the soil, marine, and plant microbiota and the important roles played by them that in turn influence them and human health and well-being

| Microbiome | Factors affecting | | Important functions | Ref. |
|---|---|---|---|---|
| | Biotic | Abiotic | | |
| Soil | Flora and fauna; human activity, e.g., farming/ urbanization | Moisture, pH, temperature, gases, minerals, C/N ratio | Nitrogen cycle, carbon cycle, etc., symbiosis in the rhizosphere, diseases | Fierer. (2017) |
| Marine | Carbon chemistry, micro- and macro-nutrients, $O_2$ content | Temperature, ocean currents, ocean stratification | Emission of dimethyl sulfide; regulation of the biological pump and carbon, nitrogen, and mineral cycling | Glockner et al. (2012) |
| Plant | Herbivory and pathogens | Temperature, pH, moisture, edaphic factors | Modulation of greenhouse gases | Turner and James (2013) |

activity along with species richness and variability of microbial community. The physicochemical properties of surface soils—e.g., pH, organic carbon, salinity, texture, and bioavailable nitrogen—show huge variations just being separated by a few microns or millimeters. Primarily the elements that affect soil formation, e.g., climate or organisms (macro- and microorganisms), significantly contribute to the above variations.

Depending on the gradient of oxygen across different layers in the soil, the variability of microorganisms can be detected establishing the fact that the soil microbiota is substantially variable and ever changing. The parameters that govern this variability may be attributed to competition between the different taxa, frequent changes in soil composition (e.g., adding fertilizers), soil disturbances due to natural (storm, flood, blizzard) or human activities (ploughing/livestock industry), type of plants growing in the area or animal habitats, etc. (Tedersoo et al. 2016) and have been tabulated in Table 6.1.

The soil microbes may be abundant, but still not every part of the soil is uniformly enveloped by microbes, illustrating that biotic and abiotic factors restrain them from dwelling on every available soil surface. Even when artificial inoculation is done, the externally introduced strains may or may not survive with equal viability, showing that an underlying interplay of various parameters eventually decides the fate of microbes and that edaphic factors do not contribute alone in this case. Mostly, a single or few dominant species (covering about 90–95% of the population) are found in every niche, and the rest are in dormant form under existing conditions at a particular time point (Ochoa-Hueso 2017; Fierer 2017). A range of stipulations influences the microbiome in terms of species richness and community composition and is continuously subject to changes.

Macro-environments and microenvironments of soil include (i) the rhizosphere (soil adjoining plant roots), (ii) soil surface layers receiving sunlight (photic layer), (iii) soil in worm casts (drilosphere), and (iv) soil proximal to water channels, (v)

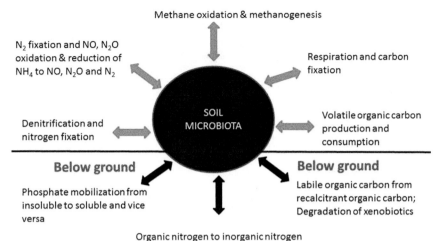

**Fig. 6.2** The biogeochemical cycles influenced by the soil microbiota

microenvironments internal to soil aggregates, and (vi) water-soil aggregates. Microbial communities change due to alterations in the abiotic parameters because of soil depth. The microorganisms found in surface soil horizons (layers) have been studied thoroughly over the years, but communities living in the litter (or O horizon) have also been studied now and were observed as unique and distinguishable from those prevailing in subsurface mineral soil horizons (A and B horizons) and deeper saprolite (C horizons).

Composition wise, there is no typical defined soil microbiota. There are several reports on some of the common biological processes occurring in the soil, e.g., methanogenesis, nitrogen fixation and nitrification, etc., and the microbial communities affecting them are well described. In spite of this knowledge, the "black box" of soil microbiota is still unknown or poorly defined for various ecosystems, e.g., tropical forests, subterraneans, and rhizospheres. Microbes also control the bio-geo cycling of various minerals and elements, e.g., S, P, and metal ions. The microbiome contributes to the amelioration of soil pollution by degrading or mineralizing various pesticides, explosives, and other xenobiotics added intentionally or accidentally by human activities (Fig. 6.2). These parameters also govern and manipulate the richness and diversity of the microbiota of a said environment.

## 6.2.2 *Plant Microbiome*

The plant microbiota has been an important area of research because it primarily determines and governs plant productivity and health (Berendsen et al. 2012), and modifications to the plant microbiome hold the promise to impact and reduce the chances of diseases and to improvise crop productivity (Bakker et al. 2012) by reducing the usage of pesticides or fertilizers, paving the way for sustainable agriculture and reducing the emission of greenhouse gases (Singh et al. 2010). Especially it is observed that, in agricultural fields, microbial denitrification and methanogenesis are stimulated by crops that play leading roles to control the emission of $N_2O$ and methane, respectively. Emission of nitrous oxide and methane depicts loss of carbon and nitrogen from the environment and leads to production of greenhouse gases. The rhizosphere is the soil mainly impacted by roots of plants due to deposition of plant exudates, mucus, and dead cells. A plant secretes a range of compounds through roots, namely, organic acids, sugars, amino acids, fatty acids, vitamins, growth factors, hormones, and antimicrobial compounds, and these vary with the type of plant and bacterial species associated with the root system. In soils where the presence of humus and organic matter is elevated, cellulose-producing microbes are commonly found dominating the microbiome. Pectin decomposition leads to release of methanol, which may be utilized as a carbon and energy source by a different group of microbes, thus affecting the structure of the microbial community. Plant cultivars influence the microbial species and strains (Inceoglu et al. 2011; Teixeira et al. 2010) dwelling any particular microbiota. The α- and β-*Proteobacteria*, *Actinobacteria*, *Firmicutes*, *Bacteroidetes*, *Planctomycetes*, *Verrucomicrobia*, and *Acidobacteria* are the most prevalent examples of rhizosphere microbiota.

Plant growth-promoting rhizobacteria are prevalent in the rhizosphere and are known to function via various mechanisms. Nitrogen-fixing bacteria are the most important examples of this category that fix nitrogen in a form that is readily available for the plant and may be included as free-living (e.g., *Azotobacter* spp.) or as symbiotic (dwelling in root nodules, e.g., *Rhizobium* spp.). Certain bacteria have the capability to solubilize minerals to release phosphorus, thereby increasing its bioavailability for plants. However, few rhizobacteria act antagonistically toward pathogens via production of antimicrobial peptides or chemicals or blocking the action of virulence factors via the action of certain effectors mediated by type 3 secretion systems (T3SSs). Actinomycetes especially provide many different types of products exhibiting antibacterial, antifungal, antiviral, nematicidal, and insecticidal properties.

Phyllosphere is the aerial surface of the plant body and is known to be comparatively poor in nutrients as compared to the rhizosphere. Leaf surfaces are colonized by up to $10^7$ microbes per $cm^2$. In such an environment, few genes show increased expression, e.g., porins, stress-related proteins, ABC (ATP-binding component) transporters, and TonB-dependent receptors, especially those belonging to *Sphingomonas* spp. Studies also showed that methylotrophs, e.g., *Methylobacterium* spp., and others dominated the community of phyllosphere microbiome (Knief et al. 2012) and could actively assimilate and metabolize methanol that is produced during degradation of pectin derived from plants. The phyllosphere, being much more subjected to fluxes in pH, temperature, moisture, and radiation throughout the diurnal cycle, experiences a greater dynamic environment compared to the rhizosphere and so do the resident microbes. Precipitation and wind especially largely contribute to the variations among inhabitant microbes and are dependent on temporal and abiotic factors. Herbivory is another prime parameter that governs phyllosphere microbiota. It is observed that the microbial species richness is higher in tropical areas that are warm and humid as compared to colder climate in temperate zones. Proteobacteria (α- and γ-) dominate the microbiota persistently; however, *Actinobacteria* and *Bacteroidetes* also prevail (Vorholt 2012).

Within the endosphere, dominance of endophytic bacteria prevails, and some of them may be pathogenic or nonpathogenic. They might be a subpopulation of the rhizosphere bacteria. It has been observed that age of plants and not biomass affects the diversity and population of inhabiting microbes. Younger plants show higher density or microbial load than older/mature ones. Specifically in case of *Herbaspirillum*, the density of epiphytic bacteria was found to be ten times higher than those of endophytes (Cavalcante et al. 2007). Root nodules of legumes populated by symbiotic bacteria may be occupied with rhizobial bacteroids up to $10^{11}$ cfu per gram of plant tissues (fresh weight). It is hypothesized that a high density of endophytic bacteria (above $10^8$ cfu per gram) may lead to elicitation of any host defense response, although, in few cases, e.g., rice and sugarcane, a wider variety and population of endophytic bacteria have been observed using culture-independent molecular methods, such as analyses of 16S rRNA and *nifH* transcripts (Fischer et al. 2012) and metagenome analyses (Sessitsch et al. 2012). Rhizobia (and other α-*Proteobacteria*) as well as β-*Proteobacteria*, γ-*Proteobacteria* and *Firmicutes* are important examples of endophytic bacteria, e.g., *Azospirillum*, *Burkholderia*, *Gluconacetobacter*, and *Herbaspirillum* spp.

### 6.2.3   Marine Microbiome

Microbes are abundantly and ubiquitously present in the marine environment, modulating key biogeochemical processes, namely, carbon and nutrient cycling. It is estimated that about $10^4$ to $10^6$ microbial cells prevail per milliliter and interestingly their biomass, along with high turnover rates in combination with considerable environmental dynamism, contributes to great diversity and species richness. It has been observed that the number of microbes in aquatic habitats is about $1.2 \times 10^{29}$ but, in the oceanic subsurface, it increases up to $3.5 \times 10^{30}$. The higher concentration of organisms in coastal waters may be due to the increased loads of organic carbon and nutrients/xenobiotics and shows higher productivities thereof. However, the deep ocean may exhibit 1–2 times less microbes comparatively (Glöckner et al. 2012). Virus population of the ocean may exceed the order of 1030. Marine environment including the seas/oceans plus coastal waters, namely, estuaries, serves as a home for various species and strains of bacteria, archaea, viruses, fungi, protists, and microalgae that transform C-, N-, P-, and S-containing molecules to simpler forms so that they are easily procured by marine plants, animals, microbes, etc. The biological pathways for metabolism of various nutrients in marine microbes thereby modulate the global biogeochemical cycles predominating upon Earth and contribute to oxygen production (indispensable for aerobic life) and carbon reduction. The interaction followed by the fine interplay of all these compounds and metabolic and biogeochemical cycles controls the dynamics of all marine microbiomes. Viruses contribute to biogeochemical cycles by lysing up to 30–50% of the microbial biomass every day. Viruses indirectly influence the abundance and diversity of host cell populations by contributing to genetic exchanges, which form the basis of evolution in host-pathogen/virus dual systems. *Cyanobacteria* also form one of the most enriched species and are significant due to its capability to fix $CO_2$ into organic matter, i.e., transformation of light to chemical energy. The carbon in fixed form may be utilized as food by higher trophic levels or remains trapped as deep sediments where it fossilizes over time to produce natural energy resources (Salazar and Sunagawa 2017). Under anaerobic conditions, methane is generated from $CO_2$ (by methanogenic bacteria) and reverted back to $CO_2$ on reversal of conditions by methanotrophic bacteria. Similarly, in biogeochemical cycles of iron, the oxidized ferric form of iron ($Fe^{3+}$) converts to reduced ferrous iron ($Fe^{2+}$) and the bacteria back-catalyze them via (abiotic) chemical processes. Due to the precipitation of iron in seawater, it is no more bioavailable to the flora and fauna, and this factor limits the growth and proliferation of microbes (Glockner et al. 2012). Sulfur is abundantly present in seawater and may be easily assimilated by microbes unlike iron. Some algae produce DMSP (dimethylsulfoniopropionate) which, when liberated, can be transformed by some bacteria and other algae to DMS (dimethyl sulfide). The DMS released is responsible for the "smell of the sea."

Very few bacteria can actually fix $N_2$ in seawater and use it for the synthesis of structural cell material and increase of biomass. The dead decaying matter stabilizes in the ocean floor where it may fossilize and get removed from the biosphere for a long period on the geological time scale (process referred to as "the biological pump").

In recent years, a new dynamics of microbiome has gained interest, implicated to stress and pathophysiology of the disease (Vayssier-Taussat et al. 2014). Therefore, a more complete understanding of the diversity of microbes that make up the human microbiome is warranted which could lead to novel therapies. The human microbiota is an assemblage of more than 100 trillion microorganisms that include bacteria, archaea, viruses, fungi, and other microeukaryotes living in symbiotic relationship within the human host and play an important role in human health and disease (Wang et al. 2017). These microorganisms of the host interact with the soil, plant, and marine microbiome and modulate their ecosystem and in turn are influenced by them that manipulate various metabolic functions through various pathways as represented in Fig. 6.1 (Jones et al. 2014).

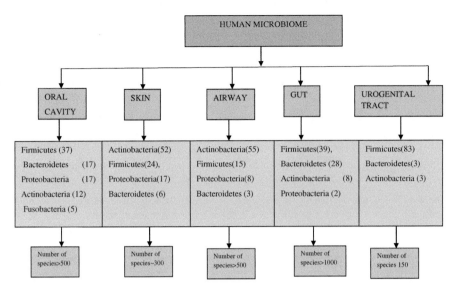

**Fig. 6.3** Human microbiota composition across five most extensive studied body sites of healthy adults. The numbers in the parenthesis represent the number of species present in each phylum

### 6.2.4   Human Microbiome

The Human Microbiome Project (HMP) conducted by the NIH in 2007 confirmed that healthy humans colonize diverse variety of microbes in its body parts. Composite microbiome that differs in each organ like oral cavity, gut, vagina, respiratory tract, skin, etc. (Argenio and Salvatore 2015) has been very elegantly summarized in various studies (Fig. 6.3), highlighting the microbial landscape across the human body which emphasizes the role of gut microbiome in healthy human adult.

## 6.3   Human Microbiome Interactions

According to genome-wide association studies (GWASs), microbiome diversity contributes in the pathogenesis of various noncommunicable diseases such as inflammatory bowel syndrome, cardiovascular diseases, diabetes, and colorectal cancer due to dysregulation in host-microbiome interactions. In general, host genetic variation plays an important role in the regulation of host-microbiome interactions. Microorganisms in the host body secrete a variety of metabolites, which

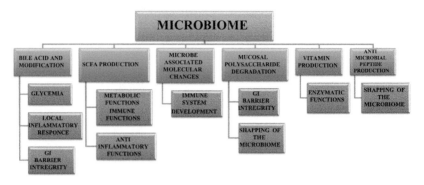

**Fig. 6.4** The human microbiota modulates several metabolic functions and intervenes in shaping the microbiome via interactions between the host and inhabiting microorganisms

alter host genetic makeup and help in modulating various functions such as immune system, enzymatic reactions, metabolic functions like bile acid, small-chain fatty acids, antimicrobial peptides synthesis, and gene expression, which in turn helps in shaping host-microbiome, anti-inflammatory response, etc. (Fig. 6.4). However, very little is known about microbiome-regulated alterations of host gene expression and vice versa (Tojo et al. 2014).

### 6.3.1 Gut Microbiome Composition and Interactions

Our knowledge regarding human intestinal microbiome is emerging, but the composition of gut microbiome and their functions in the host body is still not well defined. A variety of environmental factors like diet, antibiotic usage, mode of delivery of an infant, and their feeding and lifestyle behavior in addition to host physical status, genotype, and immune pattern can modulate the dynamics of healthy microbiome as depicted in Fig. 6.4 (Sirisinha 2016). In general, the majority of the gut microbes are harmless and beneficial to the host for maintaining normal homeostasis. However, these environmental factors may cause dysbiosis, various types of infections, and obesity-linked metabolic syndromes (e.g., diabetes and cardiovascular diseases), allergies, etc. It is also a well-established fact that excessive usage of antimicrobial peptides affects the human intestinal microbiota and decreases colonization of the beneficial habitats, which can lead to antibiotics resistance (Paul et al. 2018). In spite of the environment, host genotype plays a significant role on the composition of its microbiome and shaping an individual microbiota phenotype (Verdu et al. 2015). For example, people carrying known mutations (e.g., $NOD_2$ gene) associated with increased risk of some inflammatory bowel diseases have microbiomes that differ from those who do not have the mutation (Verdu et al. 2015).

Studies have shown that antibiotic administration impacts the human intestinal microbiota. The antimicrobial agents contribute to the decreased colonization of

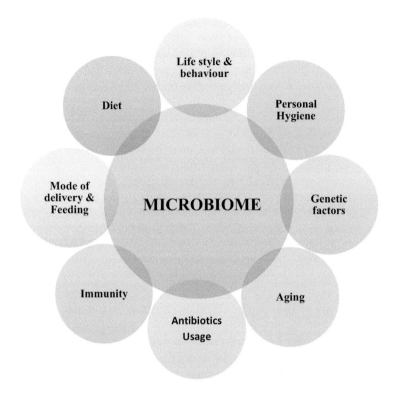

**Fig. 6.5** Influence of core microbiota by the genotype of host along with environmental factors

commensal microbiota, leading to the development of a range of diseases, including the emergence of antimicrobial resistance (Paul et al. 2018).

An elegant review by Wintermute and Silver (2010) described how host immune system responds to gut microbiota through interactions which are said to be initiated at the time of birth and continue to modulate and shape the host microbiome through cross talk transmitted via a vast array of signaling pathways. For example, microbial metabolite, such as small-chain fatty acids (SCFAs), activates entero-endocrine cells of the gut to secrete a variety of peptides that are required for digestion, lipid storage, and energy homeostasis. This is enacted through G protein-coupled receptors, like GPR41 receptor (Samuel 2008) (Fig. 6.5).

## 6.3.2 Oral Microbiome Interactions

The healthy human mouth contains hundreds of different bacterial, viral, and fungal species. Microbial flora of the oral cavity is mainly of two types: indigenous (*Streptococcus*, *Actinomyces*, and *Neisseria*) and supplemental type (*Lactobacillus* sp.). Colonization of microbes starts just after the birth (*Streptococcus salivarius*), and after 1 year it is invaded by *Streptococcus, Actinomyces, Neisseria*, and *Lactobacillus*. With due course of time, gingival crevices develop colonization along with plaque in the tooth and fissures followed by elderhood where microbiota becomes similar to childhood when all teeth are lost. Normal microbial growth depends upon multiple factors like oral anatomy, saliva, pH, diet, drugs, extraction of teeth, etc.

Host-microbiome interaction is not as well characterized in oral biofilms. However, *Veillonella* and *Streptococcus* are the most abundant genera reported which colonize in the oral cavity and live in symbiotic relationship. *Veillonella* utilize the lactate produced by the streptococci as a food source (Kuramitsu et al. 2007). In similar fashion *F. nucleatum* expresses adhesins that recognize streptococci and a lectin that interacts with *Porphyromonas gingivalis*. Similarly, the dendritic cells of oral mucosa release pro-inflammatory cytokines that activate adaptive immunity (Novak et al. 2008) and antimicrobial peptides (AMPs), such as histatins and defensins, as the first line of defense against microbes in the oral cavity and in turn interact synergistically with the microbial habitats.

### 6.3.2.1   Influence of Host Environment on the Oral Mycobiota

The oral cavity like other body parts exhibits a wide array of microbiota which may be due to alteration of many factors, including the human genetic makeup, diet, age, surroundings, and sexual behavior. Recent study advanced our knowledge that microbiota shared among partners while kissing are able to sustain in the oral cavity, transiently present in the saliva, while those on the tongue's dorsal surface are colonized for long duration (Kort et al. 2014). On the other hand, oral mycobiota is generally stable over time but varies between healthy individuals. However, the role of host factors affecting the composition of the oral mycobiota in health is still not well known. Ghannoum and co-workers suggested that gender or ethnicity may cause variation in mycobiota among individuals, but this reason is applicable when sample size is limited and there is lack of consistency across all gender groups and ethnicities (Ghannoum et al. 2010). Fungal growth like *Candida* sp. overgrowth is associated with some genetic disorders and increased risk of infection, such as candidiasis, and the autoimmune polyendocrine syndrome type I are reported (Underhill and Iliev 2014). The synergestic interaction between host factors and fungus showed that certain streptococcal species display synegism with *C. albicans* on oral mucosal or tooth surfaces. It is shown that Toll-like receptors, NOD-like receptors, and/ or C-type lectin receptors are required for fungal recognition in the oral cavity to trigger appropriate innate immune responses controlling the growth of certain fungal species while maintaining homeostasis with others. There is a growing body of evidences that bacteria and fungus are associated in the oral cavity in the same microenvironmental niches. It is evident that in immunocompromised conditions like HIV, oral fungal community shifts are accompanied by shifts in bacterial communities (Mukherjee et al. 2014). The role of individual fungi of the core oral mycobiota in the host that sustain health or promote disease remains to be elucidated.

### 6.3.3 Skin Microbiome Interactions

The skin ecosystem which is mainly predominated by bacteria, archaea, fungi, and viruses is very complex and dynamic like gut microbiome. Skin microbial communities are also much diversified and vary by site of the skin and from individual to individual. In general, skin microbiome is stable, but diverse signatures of skin microbe are found at species level when affected to some extent by environmental factors (Oh et al. 2016). The most prevalent bacterial genera of skin microbiome are *Propionibacterium, Corynebacterium, Staphylococcus,* and *Streptococcus*, present in ≥90% of the healthy subjects. These microbial habitats modulate the expression of host interleukins (IL-1a), cytokines, neuropeptides, and antimicrobial peptides (AMPs), produced by keratinocytes and sebocytes, and shape the host microbiota (Naik et al. 2012). Microbial compounds cause host immune cell activation through cross talk between immune cells and microbes (Fig. 6.6). Recent study on skin microbiota revealed that *Corynebacterium accolens* of the skin inhibit the growth of *Streptococcus pneumoniae*, a common respiratory tract pathogen. Similarly, *Mycobacterium ulcerans* releases a polyketide toxin, mycolactone, that causes Buruli ulcer in the host body (Marion et al. 2014).

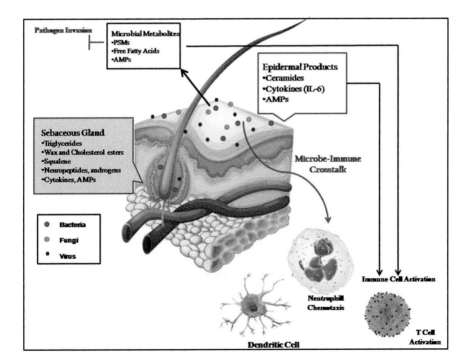

**Fig. 6.6** Distribution of various types of microbes of the skin and their interaction with secretory metabolites

The skin is inhabited by a variety of microorganisms, interacting with the host cells and modulating various cellular functions and immunities. The host in turn can influence the skin microbiota composition through cross talk and maintains the healthy status of an individual, and its disruption has been associated with disease in humans; skin microbiota differs among skin sites and among individuals.

### 6.3.4   Urine and Vaginal Microbiome Interactions

**Urine Microbiome**   The urinary microbiome of healthy individuals alters with age and has clinical outcomes but has not been well characterized. Recent study reveals that microbiota of females is more heterogeneous and alters with age groups than male urine samples and the most predominant representative phyla are *Actinobacteria* and *Bacteroidetes*. Studies reveal that bacterial composition in the urine samples of males differs from females, while *Firmicutes* are present in both male and female samples, as reported earlier (Siddiqui et al. 2011).

**Vaginal Microbiome**   A fine balance exists between the bacterial communities residing in a symbiotic relationship in human urine. *Lactobacilli* species produces lactic acid that lowers the vaginal pH to ~3.5–4.5. Vaginal environment is protected from nonindigenous and potentially harmful microorganisms through production of antimicrobial compounds like lactic acid and broad-spectrum hydrogen peroxide. *Lactobacillus* sp. are thought to play a major role in protecting the. There are various factors that have been shown to affect the vaginal microbiome which includes racial factors, hormones, use of contraceptives, sexual behavior, smoking, diet, etc. (Ravel et al. 2011; Mirmonsef et al. 2014). *Lactobacillus* typically comprise 70% or more of resident bacteria in the human vagina. The relative abundance of *Lactobacillus* and other vaginal microbial compositions is regulated by a variety of endogenous and exogenous factors. For instance, glycogen availability in the vagina is increased through estrogen-stimulated proliferation of the vaginal epithelium. Increased level of estrogen during ovulation enhances relative abundance of *Lactobacillus* spp. and causes low microbial diversity; low vaginal pH, to stabilize bacterial community (Mirmonsef et al. 2014), and, additionally, sexual contact and exposure to maternal bacteria during birth may also influence the vaginal microbiome. The seminal fluid during sexual contact transmits novel bacteria and neutralizes vaginal acidity which may impact the vaginal environment (Nunn et al. 2015). Vaginal ecosystem is also disturbed during pregnancy and has significant and long-term consequences for the offspring.

## 6.4 Microbiome in Health

### 6.4.1 Configuration of the Microbiome

The microbiome of a newly born infant starts developing even before he or she is exposed to mother's microbiota through colonization of various types of inhabitants in various parts of the body such as the skin, gut, and respiratory and urogenital tracts. These microbes mainly produce short-chain fatty acids (SCFAs) as an important source of energy (Byrne et al. 2015.), required for lipid and protein metabolism as well as in the synthesis of essential vitamins such as folates; vitamins K, $B_2$, and $B_{12}$; etc. In return, beneficial bacteria such as *Lactobacillus* and *Bifidobacterium* utilize fructo-oligosaccharides and oligosaccharides from host as energy sources for their growth.

**Child Microbiome** The first colonizers of newly born infant are facultative anaerobes. After 1 year, infants develop distinct microbial profile and, by 25 years of age, develop the characteristic of adult microbiota. Delivery mode plays an important role in developing microbiome of an individual which is distinctly different in Cesarean section and vaginally delivered child. C-section-delivered child gut is colonized with *Haemophilus* spp., *Enterobacter cancerogenus/E. hormaechei*, *Veillonella dispar/V. parvula*, and *Staphylococcus* (Bäckhed et al. 2015). These microbes retain for at least 1 year to develop immunity against infant infections. The mother's milk is known to contain more than 700 species of bacteria which play a vital role in shaping the composition of microbiome of the newly born infants.

**Microbiome at Puberty Stage** When a child enters puberty stage, she undergoes through several types of physical and physiological changes that are regulated by hormonal fluctuations. These hormonal changes depend upon the bacterial composition of vaginal microbiota, but limited studies have been done so far. Vaginal microbiota in early childhood includes aerobic, anaerobic, and enteric bacterial populations. Among them the lactic acid-producing bacteria (mainly *Lactobacillus* sp.) are the key players in maintaining homeostasis of the microbiota, as a protective response in acidic environment (Ravel et al. 2011.). However, monthly menstrual and hormonal cycles and sexual activities may cause alteration in the stability of the microbial composition (Fig. 6.7).

**Adult Microbiome** The microbial community of the gut has a great impact on the health of adult humans. For example, *Firmicutes* and *Bacteroidetes* (70–75% of total) followed by *Actinobacteria*, *Proteobacteria*, and *Verrucomicrobia* regulate vital functions of the gut metabolism (Sinha et al. 2016). Dysbiosis of the gut microbiome of the adult population leads to increased prevalence and severity of various metabolic diseases like cardiovascular disease, celiac disease, and diabetes along with increase incidences of cancer. These changes in microbiome pattern at various stages of development from pregnancy to adulthood are summarized in Fig. 6.6 and Table 6.2.

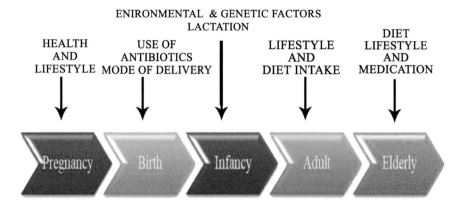

ENIRONMENTAL & GENETIC FACTORS
LACTATION

HEALTH            USE OF                         LIFESTYLE        DIET
AND             ANTIBIOTICS                        AND          LIFESTYLE
LIFESTYLE   MODE OF DELIVERY                    DIET INTAKE     MEDICATION

**Fig. 6.7** A representative schematic diagram shows epigenetic factors that modulate the microbiome of an individual starting pre-birth (pregnancy) to geriatric state of life, which ultimately affects the health

**Table 6.2** Changes in microbiome pattern at various stages of development from pregnancy to adulthood

| Stages | Predominant bacteria | Functions | References |
|---|---|---|---|
| Placental microbiome | Nonpathogenic commensal bacteria like *Firmicutes*, *Proteobacteria*, *Bacteroidetes*, and *Fusobacteria* phyla | Production of SCFAs, etc. | Aagaard et al. (2014) and Byrne et al. (2015) |
| Neonatal microbiome | *Bacteroides*, *Clostridium*, and *Bifidobacterium* spp. | Production of SCFAs, regulation of gut motility | |
| Child microbiome | Symbiotic bacteria | Lipid and protein metabolism, synthesis of vitamins and SCFAs, regulation of glucose homeostasis | |
| Puberty | *Lactobacillus crispatus*, *L. iners*, *L. gasseri*, *L. jensenii*, and, in some cases, *Streptococcus* spp. | Regulates various physiological changes in response to hormonal variation, produces various bacteriostatic and bacteriocidal compounds to protect from colonization of pathogens | |
| Adult microbiome | *Firmicutes* and *Bacteroidetes* (70–75% of total) followed by *Actinobacteria*, *Proteobacteria* and *Verrucomicrobia* | Gut metabolism, moods, and behavior | Sinha et al. (2016) |

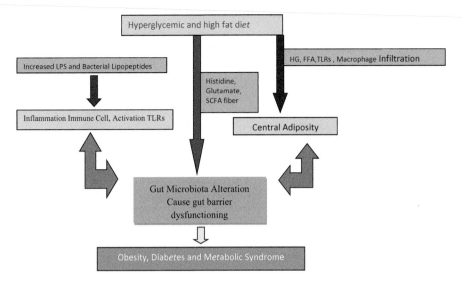

**Fig. 6.8** The diet plays a versatile role to maintain healthy microbiota. Diet rich in carbohydrates and fat leads to obesity, diabetes, and various other metabolic disorders

## 6.5 Factors Affected/Regulated by Gut Microbiome

**Diet** The composition of host microbiome is regulated by a variety of factors as discussed previously. Among them, our routine diet has been found to play a vital role in shaping the host microbiota. For example, people consuming protein and animal fat-rich diet are predominantly inhabited by *Bacteroides*, whereas those consuming carbohydrate and simple sugars possess *Prevotella* in their gut (Sirisinha 2016). Thus hyperglycemic and obese people undergo dysbiosis in their gut microbiota and are susceptible to obesity, diabetes, and other metabolic disorders as shown in Fig. 6.8 (Devaraj et al. 2013).

The molecular mechanism on how the diet regulates and maintains microbial ecosystem in the host is not known yet.

**Gut Microbiome Diversity and Host Immunity** A large number of evidences reveal that microbiota have extensive and long-lasting effects on the development and functions of both innate and adaptive immune cell populations in the gut (Sirisinha 2016). The commensal bacterium *Bacteroides thetaiotaomicron* produces AMP, C-type lectin to increase the expression of active defensins, downregulates inflammatory response by interfering with the activation of NFκB by a PPARγ-dependent pathway (Kabat et al. 2014), and can activate and regulate production of enzyme fucosyltransferases needed to maintain homeostasis. Similarly, a healthy microbiota is required for proper development of mucosal T cell subpopulations and several species of bacteria present in the gut.

**Role of Aging in Host Microbiome** The knowledge of age-related changes in the microbiota in adult is still limited. It is hypothesized that microbiota and its associated host genes that are beneficial in early life may be harmful in later stages of adulthood. For example, *H. pylori* in the early stage of life produces inflammatory responses to the organism to protect against infection, whereas in late adulthood it causes chronic ulcer, IBD, and cancer (Arnold et al. 2011). The microbiota in elder stages that shows less diversities mainly undergoes reductions in *Bifidobacterium* and *Firmicutes* and increases in *Bacteroidetes* and *Enterobacteriaceae* (Odamaki et al. 2016). Recent studies have begun to emerge that microbiome changes with aging and age-related diseases.

**Gut Microbiota: Brain Circuitry Network** The gut microbiota communicate and influence brain functions including our mood, behavioral pattern, and anxiety that depend on signals from neural, hormonal, and immune systems and from the microbiota itself. Gut microbiota can either suppress or enhance the activity of HPA (hypothalamus-pituitary axis) (Yarandi et al. 2016). There are ample evidences that bacterial species which possess receptors in the gut can respond to neurotransmitters and neuromodulators, e.g., noradrenalin, that regulate mood and stress-related behaviors (Sherwin et al. 2016). Likewise, these microbes have the ability to produce biologically active neurochemicals, e.g., serotonin (5-HT), acetylcholine, melatonin, and histamine that can modulate the activity and function of enteric nervous system (ENS) and vagus nerve (Sirisinha 2016).

## 6.5.1 Role of Gut Microbiome in Pathogenesis

Gut microbiota does play an important role in host homeostasis involving the immune system, but altered microbiota or dysbiosis can shape the disease status. Understanding these critical interactions can contribute toward designing strategies for both prevention and therapy (Wu and Lewis 2013).

(a) *Bowel diseases*: Increasing evidences show imbalances in the host-microbiota due to a bidirectional relationship between altered immune function (mucosal barrier, innate bacterial killing, or immune regulation) and altered bacterial community that leads to the onset of inflammatory bowel diseases (Knights et al. 2014). Excessive abundance of *Desulfovibrio* species is also found in ulcerative colitis.

(b) *Obesity and diabetes*: A growing body of evidences indicates that obese people are insulin resistant due to altered composition of their gut microbiota as compared to healthy individuals. For example, Tilg and Kaser have found elevated *Firmicutes*/*Bacteroidetes* ratio in insulin-resistant obese people compared to healthy people. More evidences have revealed that T2DM patients show an increased level of several opportunistic pathogens and some endotoxin-producing Gram-negative bacteria (Qin et al. 2012). Figure 6.9 represents the

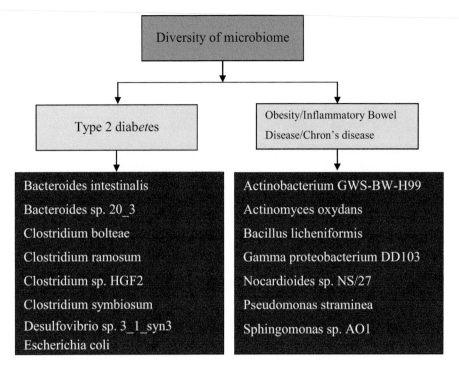

**Fig. 6.9** Diversity of microbiome in type 2 diabetes and obesity

abundance and diversity of microbiota in T2DM and obesity (Mandal et al. 2015). The potential mechanisms between the microbiota and T2DM have not been fully elucidated, and further research is needed.

(c) *Allergy*: Factors such as environment and nutrition changes can influence the inflammation-dependent diseases like allergy and asthma. Allergy mediated alteration in gut microbiota in early stages of life does lead to childhood asthma. Further the mucosal immunity, pathogen exposure, and antigen-presenting cells decode the responses of susceptibility to allergies. It has been demonstrated that Th1 cell stimulation predisposes toward allergic diseases (McLoughlin and Mills 2011). Further, use of antibiotics and dietary modifications disturbs the gut microbiome balance causing dysfunction of the immune system. These types of dysbiosis can also lead to emergence of allergic reactions. Although the association between allergy, asthma, and microbiome has only recently been under the review of research, the role of microbiota modulating the adaptive immunity could be a promising research prospect.

(d) *Colorectal cancer*: Evidence in literature suggests critical role of microbiota in progression of colon cancer. *B. fragilis* toxin (BFT)-producing strains promote colon tumorigenesis by the increased expression of STAT3 that recruits pro-inflammatory T helper lymphocytes and produce toxins for induction of TNF-α,

**Table 6.3** Differential microbiota community prevalent in colorectal cancer and healthy individual

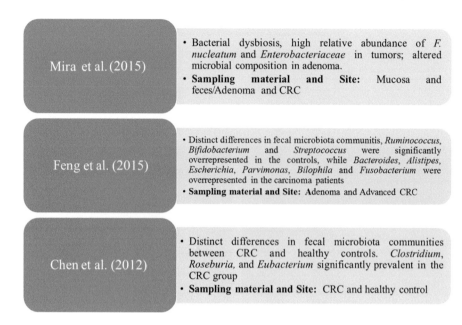

Mira et al. (2015)
- Bacterial dysbiosis, high relative abundance of *F. nucleatum* and *Enterobacteriaceae* in tumors; altered microbial composition in adenoma.
- **Sampling material and Site:** Mucosa and feces/Adenoma and CRC

Feng et al. (2015)
- Distinct differences in fecal microbiota communitis, *Ruminococcus, Bifidobacterium* and *Streptococcus* were significantly overrepresented in the controls, while *Bacteroides, Alistipes, Escherichia, Parvimonas, Bilophila* and *Fusobacterium* were overrepresented in the carcinoma patients
- **Sampling material and Site:** Adenoma and Advanced CRC

Chen et al. (2012)
- Distinct differences in fecal microbiota communities between CRC and healthy controls. *Clostridium, Roseburia,* and *Eubacterium* significantly prevalent in the CRC group
- **Sampling material and Site:** CRC and healthy control

IL-6, and COX-2 levels. They are metalloproteases and further stimulate cleavage of E-cadherin and augment β-catenin/Wnt pathway that activates CRC.

Zeller et al. (2014) explored the abundances of four most discriminative microbiota (*Fusobacterium nucleatum* subsp. *vincentii* and *animalis*, *Peptostreptococcus stomatis*, and *Porphyromonas asaccharolytica*) to correlate with CRC progression from early neoplastic growth (adenomas, stage 0/I/II) to late-stage metastasizing tumors (stage III/IV CRC patients). Several studies have clearly demonstrated the link between diet, intestinal microbiota, and the development of CRC (Table 6.3).

(e) *Oral diseases*: The microbiome found on or in the human oral cavity other than the tonsils, pharynx, and esophagus is comprised of over 600 prevalent microorganisms as represented in Fig. 6.8. These microorganisms cause a number of oral infectious diseases, including caries (tooth decay), periodontitis (gum disease), infections, etc. Moreover, there is a link between oral bacteria and a number of systemic diseases, including cardiovascular disease; and stroke, diabetes, and pneumonia are reported. In recent years, studies have shown that the risk of developing oral premalignant lesions is associated with chronic periodontitis, ultimately leading to oral squamous cell carcinoma (OSCC) (Laprise et al. 2016). Patients with chronic periodontitis often have poorly differentiated tumors within the oral cavity due to chronic inflammation and oral HPV infection (Fig. 6.10).

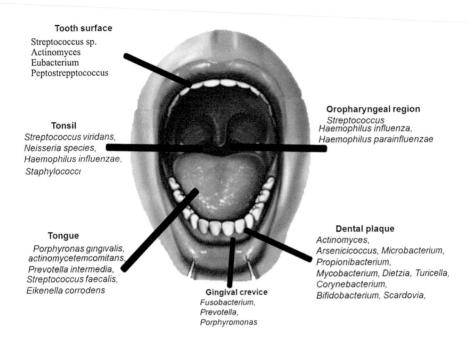

**Tooth surface**
Streptococcus sp.
Actinomyces
Eubacterium
Peptostrepptococcus

**Oropharyngeal region**
*Streptococcus*
*Haemophilus influenza,*
*Haemophilus parainfluenzae*

**Tonsil**
*Streptococcus viridans,*
*Neisseria species,*
*Haemophilus influenzae,*
*Staphylococci*

**Tongue**
*Porphyronas gingivalis,*
*actinomycetemcomitans,*
*Prevotella intermedia,*
*Streptococcus faecalis,*
*Eikenella corrodens*

**Dental plaque**
*Actinomyces,*
*Arsenicicoccus, Microbacterium,*
*Propionibacterium,*
*Mycobacterium, Dietzia, Turicella,*
*Corynebacterium,*
*Bifidobacterium, Scardovia,*

**Gingival crevice**
*Fusobacterium,*
*Prevotella,*
*Porphyromonas*

**Fig. 6.10** Diversity in microbiome distribution in various parts of the oral cavity

(f) *Skin diseases*: Human skin undergoes changes in daily life; however, healthy adults maintain their skin microbiota stable for up to 2 years (Faith et al. 2013). About ≥90% of the healthy subject's skin microbiome consists of *Propionibacterium, Corynebacterium, Staphylococcus*, and *Streptococcus* (Shi et al. 2016). Similarly, Findley et al. have found greatest diversity of fungi on the feet and bacterial species including *Staphylococcus aureus, Propionibacterium acnes*, and *Malassezia* spp., all of which are known to be beneficial for the skin but also exhibit atopic dermatitis, soft tissue infection, dandruff, etc. under certain conditions. Other well-characterized skin pathogens are papillomaviruses (causing warts), *Candida* fungal species (causing cutaneous candidiasis and diaper rash), and *Pseudomonas aeruginosa* (implicated in green nail syndrome and toe web infections). Further studies are warranted to focus on functional relevance of shifts in microbial populations that are associated with certain conditions, such as those described here.

(g) *Vaginal diseases*: Vaginal diversity in the first trimester has been found to be associated with preterm delivery risk and neonatal mortality (Haque et al. 2017). A dysbiosis in the vaginal microbiota causes yeast infections, sexually transmitted infections, urinary tract infections, and HIV infection in addition to vaginosis. A number of studies are emerging implicating the causes and consequences of shift in vaginal microbiome pattern in human system.

## 6.6    Shaping the Microbiome

### 6.6.1    Gut Microbiome

There is a growing evidence that environmental factors and high-fat diet (HFD) alter genetic composition and metabolic activity in the mammalian gut microbiome (Shen et al. 2012). This may be because of predominant changes in the relative abundance of two dominant bacterial phyla, i.e., *Firmicutes* and *Bacteroidetes*. Carmody et al. (2015) observed a significant overgrowth of *Staphylococcus aureus* and *Enterobacteriaceae* (including *E. coli*) in obese subjects. It was noticed that the animal-based diet increases the abundance of *Alistipes*, *Bilophila*, and *Bacteroides*, whereas the levels of *Roseburia*, *Eubacterium rectale*, and *Ruminococcus bromii* were found decreased. These bacteria metabolize dietary plant polysaccharides (David et al. 2014). Recent studies show the effects of diet in reshaping the gut microbiota composition. However, there is a lack of understanding about the design and implementation of dietary-based treatments that are effective in the mammalian gastrointestinal tract ecosystem.

The gut microbiota composition is partially modulated by the extracellular metabolites as bile acids (BAs). BAs are major constituents of bile, produced in the liver, and are further secreted into the duodenum which facilitate fat digestion and absorption (David et al. 2014). They are antibacterial and create strong selective forces for the intestinal microbiota. Bile acid secretion is induced by high-fat diet. Fat content regulates the time and amount of bile secretion, and it thus shapes the microbiota.

Islam et al. (2011) observed that rats fed with cholic acid (CA) have shown increased number in *Firmicutes* accompanied by decreased number of *Bacteroidetes*. This alteration of microbiota was similar to that of obesity-associated gut microbiome. This infers that bile acid contributes in shaping the obesity-associated gut microbial composition.

Genetic makeup of human beings is conditioned due to their adaptation to the environment similar to their ancestors. In mammals, bacterial diversity varies from carnivore to omnivore to herbivore. This is due to the variation in the dietary habits. Also, a large population was limited to specific areas; this created selective pressure that favored pathogens specialized in colonizing human hosts and probably produced the first wave of emerging human diseases (Blaser 2006).

Most of the developed countries successfully controlled infectious diseases during the second half of the last century, by improved sanitation and by the use of antibiotics and vaccines. At the same time, new diseases such as allergy, autoimmune disorders, and inflammatory bowel disease (IBD) emerged in both adults and in children. The microflora of the gastrointestinal tract plays a crucial role in the pathogenesis of IBD, and studies have revealed that obesity is the key factor to cause imbalance of the normal gut microbiota.

### 6.6.2   Oral Microbiome

Our oral cavity is divided into several parts, and each part is occupied with different types of habitats such as the nonkeratinized buccal mucosa, the keratinized mucosa of the tongue and gingiva, the subgingival sulcus colonized by biotic habitats, and the enamel and dental implants by abiotic microflora. Nearly, 20 billion organisms can be found in this environment representing nearly 700 different species. Recent studies indicate that variation in salivary microbiota requires a frequent bacterial exchange and is therefore most pronounced in couples with relatively high intimate kiss frequencies (Kort et al. 2014). The most abundant are *Streptococcus* and *Lactobacillus* as less abundant genus. The 16S rRNA sequencing study revealed that the bacteria and fungi mainly represent oral microbiota, and most infections are triggered by the *Candida* sp. overgrowth in a favorable host environment. However, no studies have explored the role of global fungal population shifts during oral infection. Ghannoum and colleagues have found that, as compared to the skin and other mucosal sites, the oral cavity represents significantly greater microbial diversity. These include *Candida* species (in 75% of the participants), *Cladosporium*, *Aureobasidium*, *Saccharomycetales*, *Aspergillus*, *Fusarium*, and *Cryptococcus*.

### 6.6.3   Breast Milk

Breast milk possess essential nutrients for microbial growth in the form of human milk oligosaccharides (Musilova et al. 2014), providing nearly twice the abundance of intestinal gut microbiota to breast-fed infants than their formula-fed ones. The human milk oligosaccharides are actually converted to short-chain fatty acids, which are known to promote *Bifidobacteria* and *Lactobacillus*, and thereby these species are transferred to the neonatal gut. Apart from nutrient utilization, ecological succession of the infant microbiome is believed to affect and train the naïve immune system and modulate the infant's metabolic system. Dysbiosis of normal microbiome may have downstream consequences such as autoimmune and metabolic disorders (Paul et al. 2018).

## 6.7   Future Prospective

### 6.7.1   Prebiotics

A prebiotic may be viewed as "an agent that confers some benefit to the health of the host by careful discretion" (Gibson et al. 2017). A prebiotic enriches beneficial gut microflora by altering its composition, and therefore it may be used as a therapeutic approach for treating diseases causing gut inflammation. However, gut inflammation may also be treated by monitoring the food intake or diet of the host and in turn maintains healthy microbiota to restore homeostasis, apart from using probiotics (*Saccharomyces boulardii UFMG 905* and *Bifidobacterium*). Dietary fiber-rich food enhances the growth of good bacteria (e.g., *Faecalibacterium prausnitzii, Bifidobacterium, Bacteroidetes*, namely, *Prevotella* and *Xylanibacter)* and retards harmful ones (e.g., *Firmicutes, Enterobacteriaceae*). Enhancement in the amount of *Bifidobacteria* and *Lactobacilli* is now regarded as a marker of intestinal health, and prebiotics normally promote the proliferation of these bacteria so that they can overcome the deleterious effects of harmful bacteria that cause proteolysis and/or putrefaction (Martinez 2014).

### 6.7.1.1 Beneficial Alterations in Gut Microbiota Through Prebiotics

Individuals facing the challenges of obesity reveal targeted alterations in the gut microflora phylum level as well as at genus or species level. It has been shown that lower levels of *Bifidobacteria* during the time of birth may be related to being overweight during their childhood in later years. Additionally, mothers who are overweight potentially have babies with lower levels of *Bifidobacteria* at birth, again suggesting that obesogenic microflora is "inheritable." These bacteria are comparatively less in count in patients suffering from type 2 diabetes mellitus as compared to nondiabetic patients (Wu et al. 2010). These examples suggest that, in cases of obesity and related manifestations, *Bifidobacteria* play an active and important role, thereby being identified as a model organism for the hypothesis of using prebiotics used for targeted stimulation of growth or metabolism for specific health benefits to the host. Dietary fructans, present commonly in all fruits and vegetables and are included in food products, may be utilized as a source of energy by microbes, e.g., *Bifidobacterium* spp., as it produces β-fructofuranosidase, which in turn helps them to colonize in the gut. An experiment showed a stark increase in the population of *Bifidobacterium* spp. in diet-induced obese mice and also in genetically determined obese mice once they were given a diet that was reinforced by inulin-type fructans (Dewulf et al. 2011). Surprisingly, the growth and proliferation of *Bifidobacteria* showed an inverse relation with the formation of lipids and glucose intolerance and with the lipopolysaccharide level. Alongside, the administration of prebiotic prevented overexpression of specific host genes involved in the regulation of adiposity and inflammation but regulated the gut colonization of germ-free mice.

In this aspect, it is imperative to analyze the consequence of prebiotic or probiotic advances for treatment of obesity and other metabolic disorders in human beings.

## 6.7.2  Probiotics

Probiotics are live microbial products that promote the growth of other microorganisms, when they are administered in adequate amounts beneficial to the host. Probiotics have beneficial effects in the context of different disease states as they are capable of affecting the composition and function of the host microbiome.

The supplementation of probiotics to infant diets through breast milk has been found beneficial for protection against diarrhea and rotaviral infection and is resistant to infectious gastroenteritis. Various compounds and their derivatives synthesize by gut bacteria has a tremendous impact on the physiology, immunity and disease resistance of human individuals. Examples include synthesis of B complex vitamins such as vitamin B12 (cobalamin) and vitamin B1 (thiamine) by de novo biosynthesis pathways in the gut microbiome (Saulnier et al. 2009). Probiotics containing antimicrobial agents or metabolic compounds suppress the growth of other microorganisms in the intestinal mucosa.

Probiotics also produce some secretory factors and metabolites that help in modulating the intestinal immune system and suppress the growth and function of intestinal epithelial and immune cells. For instance, the Gram-positive bacterium *L. reuteri* regulates cytokine production and signaling of immune cells. A more recent study demonstrated that a probiotic mixture of *L. acidophilus*, *L. plantarum*, *L. rhamnosus*, *Bifidobacterium breve*, *B. lactis*, *B. longum*, and *Streptococcus thermophilus* to the patients with diarrhea and irritable bowel syndrome (IBS) get symptomatic relief (Thomas and Versalovic 2010). In the context of different diseases in pediatric gastroenterology like necrotizing enterocolitis (NEC), antibiotic-associated diarrhea and colitis, acute gastroenteritis, and irritable bowel syndrome, probiotics have yielded beneficial effects. In premature infants, human milk has been effective in reducing the incidence of NEC (Sullivan et al. 2010). Probiotics have been found successful in preventing preterm delivery, including very-low-birth-weight infants. An administration of histamine 2 receptor (H2R) antagonists as acid blockers to preterm infants being shown increased incidence of NEC. The dietary amino acid L-histidine is converted to histamine by probiotic uses, to suppress inflammation by promoting H2R signaling in the intestinal mucosa (AlFaleh and Anabrees 2014).

## 6.8  Personalized Medicine

Variations in environment (epigenomic) and lifestyle for each person contribute for their specific medical condition. These factors are thus variable from one individual to another, and hence the concept of personalized therapeutic/medical options has been derived. Different diseases have different complex metabolic and microbiome pattern, which is unique for each patient. Individualized dietary composition or supplement could be designed to tackle a disease condition with minimum adverse events, thus reducing the cost of treatment burden in totality.

## 6.9    Conclusion

Microbiota plays a critical role in health of living organisms. By understanding the role of microbiome, many facets of various diseases have changed. Till date, therapeutic agents such as probiotic and prebiotic supplements, dietary interventions, and fecal microbiota transplantation have shown a potential hope in reshaping the gut microbiome. Improved understanding in the field of microbiome can develop many novel therapeutic strategies for different disease conditions.

Petrosino (2018) demonstrated the use of modified bacterium *Escherichia coli* as a transporter of an enzyme that destroys the cancer cells. Bacteria *Clostridium perfringens* that produce a protein *Clostridium perfringens* enterotoxin (CPE) target the claudin-3 and claudin-4 epithelial receptors present in breast, prostate, lung, endometrial, thyroid, and pancreatic cancer tumors (Yonogi et al. 2014). Current knowledge of such interactions is not satisfactory and represents only the tip of a very large iceberg warranting continued research in order to exploit its potential use.

Taken together the content of this chapter that illustrates the vastness of available data on diverse aspect of microbiome, its diversity, its genome variation, its secretome products, its ability to target the host immunity, and its ability to alter epigenomic factors, demands extensive further research to be able to precisely use the knowledge in the future for diagnosis and prognosis and also in targeted therapy.

**Acknowledgment** SB and SC are thankful to Mr. Sartaj Khurana, Mrs. Rizwana Mirza, Mr. Kamran Manzoor Waidha, and Mr. Rajat Gupta for helping us in the preparation of figures for this book chapter. SB would like to express his sincere gratitude to the editors for all technical supports.

## References

Aagaard, K., Ma, J., Antony, K. M., Ganu, R., Petrosino, J., & Versalovic, J. (2014). The placenta harbors a unique microbiome. *Science Translational Medicine, 6*(237), 237ra65.

Alfaleh, K., & Anabrees, J. (2014). Probiotics for prevention of necrotizing enterocolitis in preterm infants. *Evidence-Based Child Health: A Cochrane Review Journal, 9*(3), 584–671.

Arnold, I. C., Dehzad, N., Reuter, S., Martin, H., Becher, B., Taube, C., & Müller, A. (2011). Helicobacter pylori infection prevents allergic asthma in mouse models through the induction of regulatory T cells. *The Journal of Clinical Investigation, 121*(8), 3088–3093.

Bäckhed, F., Roswall, J., Peng, Y., Feng, Q., Jia, H., Kovatcheva-Datchary, P., Li, Y., Xia, Y., Xie, H., Zhong, H., & Khan, M. T. (2015). Dynamics and stabilization of the human gut microbiome during the first year of life. *Cell Host & Microbe, 17*(5), 690–703.

Bakker, M. G., Manter, D. K., Sheflin, A. M., Weir, T. L., & Vivanco, J. M. (2012). Harnessing the rhizosphere microbiome through plant breeding and agricultural management. *Plant and Soil, 360*, 1–13.

Berendsen, R. L., Pieterse, C. M. J., & Bakker, P. (2012). The rhizosphere microbiome and plant health. *Trends in Plant Science, 17*, 478–486.

Blaser, M. J. (2006). Who are we?: Indigenous microbes and the ecology of human diseases. *EMBO Reports, 7*(10), 956–960.

Blaser, M. J. (2016). Antibiotic use and its consequences for the normal microbiome. *Science, 352*(6285), 544–545. https://doi.org/10.1126/science.aad9358.

Byrne, C. S., Chambers, E. S., Morrison, D. J., & Frost, G. (2015). The role of short chain fatty acids in appetite regulation and energy homeostasis. *International Journal of Obesity, 39*(9), 1331.

Carmody, R. N., Gerber, G. K., Luevano, J. M., Jr., Gatti, D. M., Somes, L., Svenson, K. L., & Turnbaugh, P. J. (2015). Diet dominates host genotype in shaping the murine gut microbiota. *Cell Host & Microbe, 17*(1), 72–84.

Cavalcante, J. J. V., Vargas, C., Nogueira, E. M., Vinagre, F., Schwarcz, K., Baldani, J. I., Ferreira, P. C. G., & Hemerly, A. S. (2007). Members of the ethylene signalling pathway are regulated in sugarcane during the association with nitrogen-fixing endophytic bacteria. *Journal of Experimental Botany, 58*, 673–686.

Chen, W., Liu, F., Ling, Z., Tong, X., & Xiang, C. (2012). Human intestinal lumen and mucosa-associated microbiota in patients with colorectal cancer. *PLoS One, 7*(6), e39743.

D'Argenio, V., & Salvatore, F. (2015). The role of the gut microbiome in the healthy adult status. *Clinica Chimica Acta, 451*, 97–102.

David, L. A., Maurice, C. F., Carmody, R. N., Gootenberg, D. B., Button, J. E., Wolfe, B. E., Ling, A. V., Devlin, A. S., Varma, Y., Fischbach, M. A., & Biddinger, S. B. (2014). Diet rapidly and reproducibly alters the human gut microbiome. *Nature, 505*(7484), 559.

Devaraj, S., Hemarajata, P., & Versalovic, J. (2013). The human gut microbiome and body metabolism: Implications for obesity and diabetes. *Clinical Chemistry, 59*(4), 617–628.

Dewulf, E. M., et al. (2011). Inulin-type fructans with prebiotic properties counteract GPR43 over-expression and PPARγ-related adipogenesis in the white adipose tissue of high-fat diet-fed mice. *The Journal of Nutritional Biochemistry, 22*, 712–722.

Faith, J. J., Guruge, J. L., Charbonneau, M., Subramanian, S., Seedorf, H., Goodman, A. L., Clemente, J. C., Knight, R., Heath, A. C., Leibel, R. L., & Rosenbaum, M. (2013). The long-term stability of the human gut microbiota. *Science, 341*(6141), 1237439.

Feng, Q., Liang, S., Jia, H., Stadlmayr, A., Tang, L., Lan, Z., Zhang, D., Xia, H., Xu, X., Jie, Z., & Su, L. (2015). Gut microbiome development along the colorectal adenoma–carcinoma sequence. *Nature Communications, 6*, 6528.

Fierer, N. (2017). Embracing the unknown: disentangling the complexities of the soil microbiome. *Nature Reviews. Microbiology, 15*, 579–590. https://doi.org/10.1038/nrmicro.2017.87.

Fischer, D., Pfitzner, B., Schmid, M., Simoes-Araujo, J. L., Reis, V. M., Pereira, W., Ormeno-Orrillo, E., Hai, B., Hofmann, A., Schloter, M., Martinez-Romero, E., Baldani, J. I., & Hartmann, A. (2012). Molecular characterisation of the diazotrophic bacterial community in uninoculated and inoculated field-grown sugarcane(Saccharum sp.). *Plant and Soil, 356*, 83–99.

Ghannoum, M. A., Jurevic, R. J., Mukherjee, P. K., Cui, F., Sikaroodi, M., Naqvi, A., & Gillevet, P. M. (2010). Characterization of the oral fungal microbiome (mycobiome) in healthy individuals. *PLoS Pathogens, 6*(1), e1000713.

Gibson, G. R., Hutkins, R., Sanders, M. E., Prescott, S. L., Reimer, R. A., Salminen, S. J., Scott, K., Stanton, C., Swanson, K. S., Cani, P. D., & Verbeke, K. (2017). Expert consensus document: The International Scientific Association for Probiotics and Prebiotics (ISAPP) consensus statement on the definition and scope of prebiotics. *Nature Reviews. Gastroenterology & Hepatology, 14*(8), 491.

Glöckner, F. O., Stal, L. J., Sandaa, R. A., et al. (2012). Marine microbial diversity and its role in ecosystem functioning and environmental change. In J. B. Calewaert & N. McDonough (Eds.), *Marine board position paper 17* (pp. 1–84). Ostend: Marine Board-ESF.

Haque, M. M., Merchant, M., Kumar, P. N., Dutta, A., & Mande, S. S. (2017). First-trimester vaginal microbiome diversity: A potential indicator of preterm delivery risk. *Scientific Reports, 7*(1), 16145.

Inceoglu, O., Abu Al-Soud, W., Salles, J. F., Semenov, A. V., & van Elsas, J. D. (2011). Comparative analysis of bacterial communities in a potato field as determined by pyrosequencing. *PLoS One, 6*, 11.

Islam, K. S., Fukiya, S., Hagio, M., Fujii, N., Ishizuka, S., Ooka, T., Ogura, Y., Hayashi, T., & Yokota, A. (2011). Bile acid is a host factor that regulates the composition of the cecal microbiota in rats. *Gastroenterology, 141*(5), 1773–1781.

Jones, M. L., Ganopolsky, J. G., Martoni, C. J., Labbé, A., & Prakash, S. (2014). Emerging science of the human microbiome. *Gut Microbes, 5*(4), 446–457.

Kabat, A. M., Srinivasan, N., & Maloy, K. J. (2014). Modulation of immune development and function by intestinal microbiota. *Trends in Immunology, 35*(11), 507–517.

Knief, C., Delmotte, N., Chaffron, S., Stark, M., Innerebner, G., Wassmann, R., Vonmering, C., & Vorholt, J. A. (2012). Metaproteogenomic analysis of microbial communities in the phyllosphere and rhizosphere of rice. *The ISME Journal, 6*, 1378–1390.

Knights, D., Silverberg, M. S., Weersma, R. K., Gevers, D., Dijkstra, G., Huang, H., Tyler, A. D., Van Sommeren, S., Imhann, F., Stempak, J. M., & Huang, H. (2014). Complex host genetics influence the microbiome in inflammatory bowel disease. *Genome Medicine, 6*(12), 107.

Kort, R., Caspers, M., van de Graaf, A., van Egmond, W., Keijser, B., & Roeselers, G. (2014). Shaping the oral microbiota through intimate kissing. *Microbiome, 2*(1), 41.

Kuramitsu, H. K., He, X., Lux, R., Anderson, M. H., & Shi, W. (2007). Interspecies interactions within oral microbial communities. *Microbiology and Molecular Biology Reviews, 71*(4), 653–670.

Laprise, C., Shahul, H. P., Madathil, S. A., Thekkepurakkal, A. S., Castonguay, G., Varghese, I., Shiraz, S., Allison, P., Schlecht, N. F., Rousseau, M. C., & Franco, E. L. (2016). Periodontal diseases and risk of oral cancer in southern India: Results from the HeNCe life study. *International Journal of Cancer, 139*(7), 1512–1519.

Mandal, R. S., Saha, S., & Das, S. (2015). Metagenomic surveys of gut microbiota. *Genomics, Proteomics & Bioinformatics, 13*(3), 148–158.

Marion, E., Song, O. R., Christophe, T., Babonneau, J., Fenistein, D., Eyer, J., Letournel, F., Henrion, D., Clere, N., Paille, V., & Guérineau, N. C. (2014). Mycobacterial toxin induces analgesia in buruli ulcer by targeting the angiotensin pathways. *Cell, 157*(7), 1565–1576.

Martinez, F. D. (2014). The human microbiome. Early life determinant of health outcomes. *Annals of the American Thoracic Society, 11*(Suppl 1), S7–S12.

McLoughlin, R. M., & Mills, K. H. (2011). Influence of gastrointestinal commensal bacteria on the immune responses that mediate allergy and asthma. *The Journal of Allergy and Clinical Immunology, 127*(5), 1097–1107.

Mira, A., Pushker, R., & Rodríguez-Valera, F. (2006). The Neolithic revolution of bacterial genomes. *Trends in Microbiology, 14*(5), 200–206.

Mira, P. L., Cabrera, R. R., Ocon, S., Costales, P., Parra, A., Suarez, A., Moris, F., Rodrigo, L., Mira, A., & Collado, M. C. (2015 Feb 1). Microbial mucosal colonic shifts associated with the development of colorectal cancer reveal the presence of different bacterial and archaeal biomarkers. *Journal of gastroenterology., 50*(2), 167–179.

Mirmonsef, P., Hotton, A. L., Gilbert, D., Burgad, D., Landay, A., Weber, K. M., Cohen, M., Ravel, J., & Spear, G. T. (2014). Free glycogen in vaginal fluids is associated with Lactobacillus colonization and low vaginal pH. *PLoS One, 9*(7), e102467.

Mukherjee, P. K., Chandra, J., Retuerto, M., Sikaroodi, M., Brown, R. E., Jurevic, R., Salata, R. A., Lederman, M. M., Gillevet, P. M., & Ghannoum, M. A. (2014). Oral mycobiome analysis of HIV-infected patients: Identification of Pichia as an antagonist of opportunistic fungi. *PLoS Pathogens, 10*(3), e1003996.

Musilova, S., Rada, V., Vlkova, E., & Bunesova, V. (2014). Beneficial effects of human milk oligosaccharides on gut microbiota. *Beneficial Microbes, 5*(3), 273–283.

Naik, S., Bouladoux, N., Wilhelm, C., Molloy, M. J., Salcedo, R., Kastenmuller, W., Deming, C., Quinones, M., Koo, L., Conlan, S., & Spencer, S. (2012). Compartmentalized control of skin immunity by resident commensals. *Science, 337*(6098), 1115–1119.

Novak, N., Haberstok, J., Bieber, T., & Allam, J. P. (2008). The immune privilege of the oral mucosa. *Trends in Molecular Medicine, 14*(5), 191–198.

Nunn, K. L., Wang, Y. Y., Harit, D., Humphrys, M. S., Ma, B., Cone, R., Ravel, J., & Lai, S. K. (2015). Enhanced trapping of HIV-1 by human cervicovaginal mucus is associated with Lactobacillus crispatus-dominant microbiota. *MBio, 6*(5), e01084–e01015.

Ochoa-Hueso, R. (2017). Global change and the soil microbiome: a human-health perspective. *Frontiers in Ecology and Evolution, 5*, 71. https://doi.org/10.3389/fevo.2017.00071.

Odamaki, T., Kato, K., Sugahara, H., Hashikura, N., Takahashi, S., Xiao, J. Z., Abe, F., & Osawa, R. (2016). Age-related changes in gut microbiota composition from newborn to centenarian: A cross-sectional study. *BMC Microbiology, 16*(1), 90.

Oh, J., Byrd, A. L., Park, M., Kong, H. H., & Segre, J. A. (2016). NISC Comparative Sequencing Program. Temporal stability of the human skin microbiome. *Cell, 165*(4), 854–866.

Paul, D., Manna, S., & Mandal, S. M. (2018). Antibiotics associated disorders and post-biotics induced rescue in gut health. *Current Pharmaceutical Design, 24*(7), 821–829.

Petrosino, J. F. (2018 Dec). The microbiome in precision medicine: The way forward. *Genome Medicine, 10*(1), 12.

Qin, J., et al. (2012). A metagenome-wide association study of gut microbiota in type 2 diabetes. *Nature, 490*(7418), 55–60.

Ravel, J., Gajer, P., Abdo, Z., Schneider, G. M., Koenig, S. S., McCulle, S. L., Karlebach, S., Gorle, R., Russell, J., Tacket, C. O., & Brotman, R. M. (2011). Vaginal microbiome of reproductive-age women. *Proceedings of the National Academy of Sciences, 108*(Supplement 1), 4680–4687.

Salazar, G., & Sunagawa, S. (2017). Marine microbial diversity. *Current Biology, 27*, R431–R510.

Samuel, B. S. (2008). Effects of the gut microbiota on host adiposity are modulated by the short-chain fatty-acid binding G protein-coupled receptor, Gpr41. *Proceedings of the National Academy of Sciences of the United States of America, 105*(43), 16767–16772.

Saulnier, D. M., Spinler, J. K., Gibson, G. R., & Versalovic, J. (2009). Mechanisms of probiosis and prebiosis: Considerations for enhanced functional foods. *Current Opinion in Biotechnology, 20*(2), 135–141.

Serna-Chavez, H. M., Fierer, N., & van Bodegom, P. M. (2013). Global drivers and patterns of microbial abundance in soil. *Global Ecology and Biogeography, 22*, 1162–1172.

Sessitsch, A., Hardoim, P., Doring, J., Weilharter, A., Krause, A., Woyke, T., Mitter, B., Hauberg-Lotte, L., Friedrich, F., Rahalkar, M., Hurek, T., Sarkar, A., Bodrossy, L., van Overbeek, L., Brar, D., van Elsas, J. D., & Reinhold-Hurek, B. (2012). Functional characteristics of an endophyte community colonizing rice roots as revealed by metagenomic analysis. *Molecular Plant-Microbe Interactions, 25*, 28–36.

Shen, Q., Zhao, L., & Tuohy, K. M. (2012). High-level dietary fibre up-regulates colonic fermentation and relative abundance of saccharolytic bacteria within the human faecal microbiota in vitro. *European Journal of Nutrition, 51*, 693–705.

Sherwin, E., Rea, K., Dinan, T. G., & Cryan, J. F. (2016). A gut (microbiome) feeling about the brain. *Current Opinion in Gastroenterology, 32*(2), 96–102.

Siddiqui, H., Nederbragt, A. J., Lagesen, K., Jeansson, S. L., & Jakobsen, K. S. (2011). Assessing diversity of the female urine microbiota by high throughput sequencing of 16S rDNA amplicons. *BMC Microbiology, 11*(1), 244.

Siddiqui, H., Nederbragt, A. J., Lagesen, K., Jeansson, S. L., & Jakobsen, K. S. (2011). Assessing diversity of the female urine microbiota by high throughput sequencing of 16S rDNA amplicons. *BMC Microbiol, 11*(1), 244.

Singh, B. K., Bardgett, R. D., Smith, P., & Reay, D. S. (2010). Microorganisms and climate change: Terrestrial feedbacks and mitigation options. *Nature Reviews. Microbiology, 8*, 779–790.

Sinha, R., Ahn, J., Sampson, J. N., Shi, J., Yu, G., Xiong, X., Hayes, R. B., & Goedert, J. J. (2016). Fecal microbiota, fecal metabolome, and colorectal cancer interrelations. *PLoS One, 11*(3), e0152126.

Sirisinha, S. (2016). The potential impact of gut microbiota on your health: Current status and future challenges. *Asian Pacific Journal of Allergy and Immunology, 34*(4), 249–264.

Sullivan, S., Schanler, R. J., Kim, J. H., Patel, A. L., Trawöger, R., Kiechl-Kohlendorfer, U., Chan, G. M., Blanco, C. L., Abrams, S., Cotten, C. M., & Laroia, N. (2010). An exclusively human milk-based diet is associated with a lower rate of necrotizing enterocolitis than a diet of human milk and bovine milk-based products. *The Journal of Pediatrics, 156*(4), 562–567.

Tedersoo, L., Bahram, M., Cajthaml, T., Põlme, S., Hiiesalu, I., Anslan, S., Harend, H., Buegger, F., Pritsch, K., Koricheva, J., & Abarenkov, K. (2016). Tree diversity and species identity effects on soil fungi, protists and animals are context dependent. *The ISME Journal, 10*, 346–362.

Teixeira, L., Peixoto, R. S., Cury, J. C., Sul, W. J., Pellizari, V. H., Tiedje, J., & Rosado, A. S. (2010). Bacterial diversity in rhizosphere soil from Antarctic vascular plants of Admiralty Bay, maritime Antarctica. *The ISME Journal, 4*, 989–1001.

Thomas, C. M., & Versalovic, J. (2010). Probiotics-host communication: Modulation of signaling pathways in the intestine. *Gut Microbes, 1*(3), 148–163.

Tojo, R., Suárez, A., Clemente, M. G., de los Reyes-Gavilán, C. G., Margolles, A., Gueimonde, M., & Ruas-Madiedo, P. (2014). Intestinal microbiota in health and disease: Role of bifidobacteria in gut homeostasis. *World Journal of Gastroenterology, 20*(41), 15163.

Turner, T. R., & James, E. K. (2013). Poole PS. *Genome Biology, 14*, 209 http://genomebiology.com/2013/14/6/209.

Underhill, D. M., & Iliev, I. D. (2014). The mycobiota: Interactions between commensal fungi and the host immune system. *Nature Reviews. Immunology, 14*, 405–416.

Vayssier-Taussat, M., Albina, E., Citti, C., Cosson, J. F., Jacques, M. A., Lebrun, M. H., Le Loir, Y., Ogliastro, M., Petit, M. A., Roumagnac, P., & Candresse, T. (2014). Shifting the paradigm from pathogens to pathobiome: New concepts in the light of meta-omics. *Frontiers in Cellular and Infection Microbiology, 4*, 29.

Verdu, E. F., Gallipeau, H. J., & Jabri, B. (2015). Novel players in coeliac disease pathogenesis: Role of the gut microbiota. *Nature Reviews. Gastroenterology & Hepatology, 12*, 497–506.

Vorholt, J. A. (2012). Microbial life in the phyllosphere. *Nature Reviews. Microbiology, 10*, 828–840.

Wang, B., Yao, M., Lv, L., Ling, Z., & Li, L. (2017). The human microbiota in health and disease. *Engineering, 3*(1), 71–82.

Wintermute, E. H., & Silver, P. A. (2010). Emergent cooperation in microbial metabolism. *Molecular Systems Biology, 6*(1), 407.

Wu, G. D., & Lewis, J. D. (2013). Analysis of the human gut microbiome and association with disease. *Clinical Gastroenterology and Hepatology, 11*(7), 774–777.

Wu, X., et al. (2010). Molecular characterisation of the faecal microbiota in patients with type II diabetes. *Current Microbiology, 61*, 69–78.

Yarandi, S. S., Peterson, D. A., Treisman, G. J., Moran, T. H., & Pasricha, P. J. (2016 Apr). Modulatory effects of gut microbiota on the central nervous system: How gut could play a role in neuropsychiatric health and diseases. *Journal of Neurogastroenterology and Motility, 22*(2), 201.

Yonogi, S., Matsuda, S., Kawai, T., Yoda, T., Harada, T., Kumeda, Y., Gotoh, K., Hiyoshi, H., Nakamura, S., Kodama, T., & Iida, T. (2014). BEC, a novel enterotoxin of Clostridium perfringens found in human clinical isolates from acute gastroenteritis outbreaks. *Infection and Immunity, 82*(6), 2390–2399.

Zeller, G., Tap, J., Voigt, A. Y., Sunagawa, S., Kultima, J. R., Costea, P. I., Amiot, A., Böhm, J., Brunetti, F., Habermann, N., & Hercog, R. (2014). Potential of fecal microbiota for early-stage detection of colorectal cancer. *Molecular Systems Biology, 10*(11), 766.

# Chapter 7
# Bioinformatics Resources

**Neetu Jabalia**

**Abstract** Bioinformatics is an interdisciplinary research area at the interface between computer sciences and biological sciences. One of the goals of this chapter is to give a predominant perception of living cell and its functions at the molecular level using bioinformatics approaches including databases, tools, visualization, and data analysis. These approaches are implied at various levels such as metabolites, transcripts, and proteins. Therefore, the major focus of the present chapter will include many applications of bioinformatics in the area of genomics, proteomics, transcriptome, and metabolomics. Automated data-gathering tools are used for clustering and analysis of experimentally derived genomic data. Different in silico tools are used with implications both in structural and functional genomics. The chapter gives a detailed overview of the significant tools used for structural genomics such as TIGR assembler, VecScreen, EULER, Phred, and Phrap. Glimpses of comparative genomics approaches, namely, MAVID, LAGAN, BLASTZ, PipMaker, CoreGenes, and GeneOrder, are elaborated with a focus on gene functions at the whole genome level. A snapshot of high-throughput approaches using ESTs includes UniGene, TIGR Gene Indices, and SAGE (SAGEmap, SAGE Geneie, SAGExProfiler) and microarray-based approaches (SOTA, TIGR Tm4, Array Designer 2, Array mining) facilitates in understanding the interaction of genes and their regulations. The central dogma of life is incomplete without an understanding of each level spanning from genomics to proteomics. Thus, an exhaustive proteome analysis will immensely help in the elucidation of cellular functions. The latter dimension is covered by protein expression analysis tools such as Melanie, SWISS-2DPAGE, Comp 2D gel, protein identification through database searching (Mascot, ProFound, PepIdent), posttranslational modifications (AutoMotif, FindMod), protein sorting (TargetP, SignalO, PSORT), and protein–protein interactions (STRING, APID, InterPreTS). The last section describes the databases and mining software used for data integration, data interpretation, and metabolomics data in system biology. A brief explanation about commercial software, namely, ChromaTOF,

N. Jabalia (✉)
Amity Institute of Biotechnology, Amity University, Noida, Uttar Pradesh, India
e-mail: njabalia@amity.edu

© Springer Nature Singapore Pte Ltd. 2018
P. Arivaradarajan, G. Misra (eds.), *Omics Approaches, Technologies And Applications*, https://doi.org/10.1007/978-981-13-2925-8_7

GeneSpring MS, MarkerView, Mass Frontier, MarkerLynx, and complex LC/MS data analysis (BLSOM, Chrompare, MathDAMP), will help the readers to effectively use the information for their research endeavors.

**Keywords** Genomics · Proteomics · Transcriptomics · Metabolomics · Bioinformatics

## 7.1 Introduction

Over the decades, the significant progressions in omics advances have empowered a high-throughput observation of an assortment of molecular and organismal processes, widely applied to identify biological variants (e.g., biomarkers, proteins, or nucleotide sequences), to characterize complex biochemical systems, and to study pathophysiological processes. While many omics platforms target comprehensive analysis of genes (genomics), mRNA (transcriptomics), proteins (proteomics), and metabolites (metabolomics) (Gracie et al. 2011), challenges still remain within information.

A genome can be described at the highest resolution by a complete genome sequence. Genomic studies are characterized by simultaneous identification of a huge number of genes using automated information gathering tools. A range of topics from genome mapping, sequencing, and functional genomic analysis to comparative genomic analysis are grouped under genomics. Protein expression analysis at the proteome level promises more accurate elucidation of cellular functions. It encompasses a range of activities including large-scale identification, quantification, determination of their localization, modifications, interactions, and functions of proteins. This chapter covers the major topics in proteomics such as analysis of protein expression, posttranslational modifications, protein sorting, and protein–protein interactions with an emphasis on bioinformatics applications.

Transcriptome analysis is an expression identification of RNA molecules produced by cells, facilitates our understanding of how sets of genes work together to form metabolic, regulatory, and signaling pathways within the cell. It reveals patterns of coexpressed and coregulated genes and allows determination of the functions of genes that were previously uncharacterized. This chapter mainly discusses the bioinformatics aspect of the transcriptome analysis that can be conducted using either sequence- or microarray-based approaches.

Metabolomics is another omics approach majorly applied to analyze small molecules and biochemical intermediates (metabolites); moreover, it is applied to identify the biomarker and treatment efficacy monitoring in cancer or type I diabetes (Friedrich 2012; Fahrmann et al. 2015; Wikoff et al. 2015). This chapter focuses on various tools for the association of metabolomics with genomics, transcriptomics, and proteomics data (Fig. 7.1).

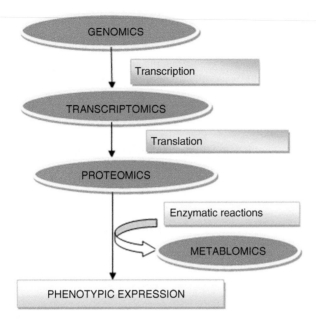

**Fig. 7.1** Flowchart of omics technologies

## 7.2 Bioinformatics Approaches in Genomics

For characterizing genome sequence information (structural, functional, organizational), the recent genomic advancements joined the genome sequencing technologies and various in silico approaches. A consortium, GENCODE, has been developed which is majorly applied to analyze gene feature expressed in human genome (Harrow et al. 2012; Venter et al. 2015). Genome assemblers were developed for sequencing entire genomes, which was the result of close interaction between biologist and computational scientists (Simpson and Pop 2015).

### 7.2.1 Structural Genomics

Various computation approaches have been established for structural genomics for improved quality and better integrity of the associated genome information. TIGR assembler, a tool to assemble large shotgun sequencing projects, used for assembling large shotgun DNA sequences. EULER is a program for de novo assembly of reads. Phred and Phrap are a base-calling program for nucleotide sequence traces and a leading program for DNA sequence assembly, respectively.

### 7.2.1.1    Genome Assembly

Computational biologist experts provide grid system to biologist so that, before assembling the genome, quality sequencing information, overall GC content, and duplication reads should be identified; therefore, different bioinformatics resources are applied for genome assembly analysis such as the following.

**FastQC**  FastQC trim data (low quality) and reads resulting from PCR duplications can be performed with a variety of different software and scripts that simultaneously give statistics which is a beneficial beginning point available at http://www. bioinformatics.babraham.ac.uk/projects/fastqc.

**In Silico Whole Genome Sequencer (iWGS) and Analyzer**  Developed for computational biologist which aims to analyze de novo genome sequencing (Zhou et al. 2016) available at https://github.com/zhouxiaofan1983/iWGS.

**CLC Genomics Workbench**  Is developed by scientists by incorporating algorithms to analyze and visualize next-generation sequencing information available at http://www.clcbio.com/products/clc-genomics-workbench/.

**RepeatMasker**  Commonly used tool in genomics analysis (e.g., classify or identify repeats in sequence). It is freely available at http://www.repeatmasker.org/.

**EGassember**  Applied for genome sequence alignment and merges the fragments in order to generate the original segment or gene (https://www.genome.jp/tools/egassembler/).

**PBSIM**  PacBio reads simulator – toward accurate genome assembly (Ono et al. 2013) (http://code.google.com/p/pbsim/).

**Tracembler**  Is a database applied for sequence assembly and identification of genes of interest; program is available at http://www.plantgdb.org/tool/tracembler/.

**UCSC Genome Bioinformatics**  The University of California Santa Cruz Genome Bioinformatics (http://genome.ucsc.edu) provides various genome analysis tools (Kuhn et al. 2009).

**SOAPaligner/Soap2**  Is a member of Short Oligonucleotide Analysis Package (SOAP) (http://soap.genomics.org.cn/soapaligner.html). It is an updated version of SOAP software for short oligonucleotide alignment.

### 7.2.1.2 Genome Annotation

Before the assembled sequence is deposited into a database, it has to be analyzed for useful biological features. The genome annotation process provides comments for the features. This involves two steps: gene prediction and functional assignment. Gene ontology (GO) project has been developed which uses a limited vocabulary to describe molecular functions, biological processes, and cellular components. A GO description of a protein provides three sets of information: biological process, cellular component, and molecular function, each of which uses a unique set of non-overlapping vocabularies. The standardization of the names, activities, and associated pathways provides consistency in describing overall protein functions and facilitates grouping of proteins of related functions. Using GO, a genome annotator can assign functional properties of a gene product at different hierarchical levels, depending on how much is known about the gene product. At present, the GO databases have been developed for a number of model organisms by an international consortium, in which each gene is associated with a hierarchy of GO terms.

### 7.2.1.3 Comparative Genomics

Comparison of whole genomes from different organisms is comparative genomics, which includes comparison of gene number, gene location, and gene content from these genomes. The comparison helps to reveal the extent of conservation among genomes, which will provide insights into the mechanism of genome evolution and gene transfer among genomes. It helps to understand the pattern of acquisition of foreign genes through lateral gene transfer. Various tools for comparative genomics are as follows.

**MUMmer** Maximal unique match (MUMmer) is a free UNIX program from TIGR for alignment of two entire genome sequences and comparison of the locations of orthologs (http://mummer.sourceforge.net/). The program is essentially a modified BLAST, which, in the seeding step, finds the longest approximate matches that include mismatches instead of finding exact k-mer matches as in regular BLAST. The result of the alignment of whole genomes is shown as a dot plot with lines of connected dots to indicate collinearity of genes. It is optimized for pairwise comparison of closely related microbial genomes.

**BLASTZ** BLASTZ is a UNIX program modified from BLAST to do pairwise alignment of very large genomic DNA sequences (http://biosrv.cab.unina.it/blastz-web/). The modified BLAST program first masks repetitive sequences and searches for closely matched "words," which are defined as 12 identical matches within a stretch of 19 nucleotides. The words serve as seeds for extension of alignment in both directions until the scores drop below a certain threshold. Nearby aligned regions are joined by using a weighted scheme that employs a unique gap penalty scheme that tolerates minor variations such as transitions in the seeding step of the alignment construction to increase its sensitivity.

**LAGAN** Limited Area Global Alignment of Nucleotides (LAGAN) is a web-based program designed for pairwise alignment of large genomes (http://lagan.stanford.edu). It first finds anchors between two genomic sequences using an algorithm that identifies short, exactly matching words. Regions that have high density of words are selected as anchors. The alignments around the anchors are built using the Needleman–Wunsch global alignment algorithm. The unique feature of this program is that it is able to take into account degeneracy of the genetic codes and is therefore able to handle more distantly related genomes. Multi-LAGAN, an extension of LAGAN, available from the same website, performs multiple alignments of genomes using a progressive approach similar to that used in Clustal.

**PipMaker** PipMaker is a web server using the BLASTZ heuristic method to find similar regions in two DNA sequences. It produces a textual output of the alignment result and also a graphical output that presents the alignment as a percent identity plot as well as a dot plot (http://pipmaker.bx.psu.edu/pipmaker/). For comparing multiple genomes, MultiPipMaker is available from the same site.

**MAVID** MAVID is a web-based program for aligning multiple large DNA sequences, based on a progressive alignment algorithm similar to Clustal (http://baboon.math.berkeley.edu/mavid/). It produces a neighbor-joining tree as a guide tree. The sequences are aligned recursively using a heuristic pairwise alignment program called AVID. AVID works by first selecting anchors using the Smith–Waterman algorithm and then building alignments for the sequences between nearby anchors.

**GenomeVISTA** Is a database searching program that searches against the human, mouse, rat, or *Drosophila* genomes using a large piece of DNA as query (http://genome.lbl.gov/vista/index.shtml). It uses a program called BLAT to find anchors and extends the alignment from the anchors using AVID. It produces a graphical output that shows the sequence percent identity.

**CoreGenes** CoreGenes is a web-based program that determines a core set of genes based on comparison of four small genomes (http://pumpkins.ib3.gmu.edu:8080/CoreGenes). The user supplies NCBI accession numbers for the genomes of interest, and the program performs an iterative BLAST comparison to find orthologous genes by using one genome as reference and another as query.

**ACT**  Artemis Comparison Tool (ACT) is a pairwise genomic DNA sequence comparison program (written in Java and run on UNIX, Macintosh, and Windows) for detecting gene insertions and deletions among related genomes (https://www.sanger.ac.uk/science/tools/artemis-comparison-tool-act).

**SWAAP**  SWAAP is a Windows program that is able to distinguish coding versus noncoding regions and measure GC skews and oligonucleotide frequencies in a genomic sequence (http://www.bacteriamuseum.org/SWAAP/SwaapPage.html).

**GeneOrder**  When the order of a number of linked genes is conserved between genomes, it is called synteny. Order conservation is in fact rarely observed among divergent species; therefore, comparison of syntenic relationships is normally carried out between relatively close lineages. However, if syntenic relationships for certain genes are indeed observed among divergent prokaryotes, they often provide important clues to functional relationships of the genes of interest. GeneOrder (http://pumpkins.ib3.gmu.edu:8080/geneorder/) is a web-based program that allows direct comparison of a pair of genomic sequences of less than 2 Mb.

## 7.2.2  Functional Genomics

The functional genomics attempts to describe the functions and interactions of genes and proteins by making use of genome-wide approaches, in contrast to the gene-by-gene approach of classical molecular biology techniques (Bunnik and Roch 2013). It includes information derived from the many biological processes, i.e., both coding and noncoding transcription, protein translation, interactions (protein–protein or protein–DNA), and gene expression. In short, functional genomics provides insight into the biological functions of the whole genome through automated high-throughput expression analysis.

## 7.3  Bioinformatics Approaches in Proteomics

Proteomics majorly defines all the proteins present in cell, a tissue, or an organism. Computational biologist developed high-throughput approaches where large-scale analysis of proteins can be done such as functional and structural and interaction or localization. There are other programs developed for protein identification: ELISAs, 2D gel electrophoresis, protein microarrays, and mass spectrometry (Table 7.1).

**Table 7.1** Tools/servers based on proteomics

| S. No | Proteomics approach | Tool name | Description | Link |
|---|---|---|---|---|
| 1 | ELISA | ELISA-BASE | ELISA-BASE is database for capturing, organizing, and analyzing enzyme-linked immunosorbent assay microarray data | http://www.pnl.gov/statistics/ProMAT/ELISA-BASE.stm |
| 2 | 2D gel electrophoresis | Melanie | Melanie is a comprehensive software for visualization, matching, detection, quantitation, and analysis of 2D gel electrophoresis images | http://2d-gel-analysis.com/ |
| | | SWISS-2DPAGE | SWISS-2DPAGE contains information on proteins analysis on many 2D PAGE and SDS-PAGE reference maps | https://world-2dpage.expasy.org/swiss-2dpage/protein/ac=P02570 |
| | | CAROL | The CAROL software has been integrated into the gel analysis software package PDQuest by Bio-Rad | http://gelmatching.inf.fu-berlin.de/Carol.html |
| 3 | NMR | Mascot | Mascot server is developed for protein identification using mass spectrometry data | http://www.matrixscience.com/ |
| | | ProFound | ProFound applied scoring system using additional information such as peptide mass fingerprinting algorithms, in the sample protein | https://omictools.com/profound-tool |
| | | PepIdent | PepIdent is a program used in characterizing isoelectric point, molecular weight, and peptide mass fingerprinting data | https://iop.vast.ac.vn/theor/conferences/smp/1st/kaminuma/ExPASy/peptident.html |
| 4 | LC-MS | BLSOM | Batch-learning self-organizing map (BLSOM) for phylogenetic classification of metagenomic sequences obtained from mixed genomes of microorganisms | http://bioinfo.ie.niigata-u.ac.jp/?BLSOM |
| | | Chrompare | Chrompare is a software for analysis of chromatographic data. It allows automated univariate peak-by-peak comparison of complex chromatograms | http://www.chrompare.com/ |
| | | MathDAMP | MathDAMP helps to visualize the variation between metabolite profiles acquired by hyphenated MS (mass spectrometry) approaches | http://mathdamp.iab.keio.ac.jp/ |

## 7.3.1  Posttranslational Modifications

Another important aspect of the proteome analysis concerns posttranslational modifications. To assume biological activity, many nascent polypeptides have to be covalently modified before or after the folding process. This is especially true in eukaryotic cells where most modifications take place in the endoplasmic reticulum and the Golgi apparatus. The modifications include proteolytic cleavage; formation of disulfide bonds; addition of phosphoryl, methyl, acetyl, or other groups onto certain amino acid residues; or attachment of oligosaccharides or prosthetic groups to create mature proteins. Posttranslational modifications have a great impact on protein function by altering the size, hydrophobicity, and overall conformation of the proteins. The modifications can directly influence protein–protein interactions and distribution of proteins to different subcellular locations. It is therefore important to use bioinformatics tools to predict sites for posttranslational modifications based on specific protein sequences such as the following.

**AutoMotif** Is a web server predicting protein sequence motifs using the SVM approach. In this process, the query sequence is chopped up into a number of overlapping fragments, which are fed into different kernels (http://automotif.bioinfo.pl/).

**ExPASy** **Ex**pert **P**rotein **A**nalysis **Sy**stem (ExPASy) contains a number of programs to determine posttranslational modifications based on MS molecular mass data (www.expasy.ch/tools). FindMod is a subprogram that uses experimentally determined peptide fingerprint information to compare the masses of the peptide fragments with those of theoretical peptides. If a difference is found, it predicts a particular type of modification based on a set of predefined rules. It can predict 28 types of modifications, including methylation, phosphorylation, lipidation, and sulfation. GlyMod is a subprogram that specializes in glycosylation determination based on the difference in mass between experimentally determined peptides and theoretical ones.

**RESID** Is an independent posttranslational modification database listing 283 types of known modifications (http://pir.georgetown.edu/pirwww/search/textresid.html).

## 7.3.2  Prediction of Disulfide Bridges

A disulfide bridge is a unique type of posttranslational modification in which covalent bonds are formed between cysteine residues. Disulfide bonds are important for maintaining the stability of certain types of proteins. The following program is one of the publicly available programs specialized in disulfide prediction.

**Cysteine** Is a web server that predicts the disulfide bonding states of cysteine residues in a protein sequence by building profiles based on multiple sequence alignment information (http://cassandra.dsi.unifi.it/cysteines/).

### 7.3.3   Protein Sorting

Subcellular localization is an integral part of protein functionality. The study of the mechanism of protein trafficking and subcellular localization is the field of protein sorting that has become one of the central themes in modern cell biology. There are many proteins which exhibit functions only after being transported to certain compartments of the cell. For protein sorting, various computational servers are available.

**TargetP** TargetP predicts the subcellular location of eukaryotic proteins (http://www.cbs.dtu.dk/services/TargetP/).

**SignalP** SignalP predicts the presence and location of signal peptide cleavage sites in amino acid sequences from different organisms (http://www.cbs.dtu.dk/services/SignalP/).

**Psort** Psort is a tool used for analysis of localization of protein sites in cells (https://psort.hgc.jp/).

### 7.3.4   Protein–Protein Interactions

In general, proteins have to interact with each other to carry out biochemical functions. Thus, mapping out protein–protein interactions is another important aspect of proteomics. Inter-protein interactions include strong interactions that allow formation of stable complexes and weaker ones that exist transiently. Proteins involved in forming complexes are generally more tightly coregulated in expression than those involved in transient interactions (REF). Decades of research on protein biochemistry and molecular biology have accumulated tremendous amount of data related to protein–protein interactions, which allow the extraction of some general rules governing these interactions. These rules have facilitated the development of algorithms for automated analysis of protein–protein interactions which are as follows.

**Search Tool for the Retrieval of Interacting Genes (STRING)** A web server used to identify gene and protein functional associations based on combined evidence of gene linkage, gene fusion, and phylogenetic profiles (http://www.bork.embl-heidelberg.de/STRING/).

**APID**  **A**gile **P**rotein **I**nteractomes **D**ata server (APID) is a web server that gives a collection of protein interactomes (>400 organisms) based on the integration of known experimentally validated protein–protein physical interactions (http:// cicblade.dep.usal.es:8080/APID/init.action).

**InterPreTS**  Interaction Prediction through Tertiary Structure (InterPreTS) is used for predicting protein–protein interactions and available at http://www.russelllab. org/cgi-bin/tools/interprets.pl.

## 7.3.5   Predicting Interactions Based on Sequence Homology

If a pair of proteins from one proteome is known to interact, their conserved homo-logues in another proteome are likely to have similar interactions. This method relies on the correct identification of orthologs and the use of existing protein inter-action databases. The interaction predicting web servers are as follows.

**InterPreTS**  Is a web server that has a built-in database for interacting domains based on known three-dimensional protein structures (www.russell.embl-heidel-berg.de/people/patrick/interprets/interprets.html).

**IPPRED**  Is a similar web-based program that allows the user to submit multiple protein sequences (http://cbi.labri.fr/outils/ippred/IS part simple.php).

## 7.3.6   Predicting Interactions Based on Phylogenetic Information

Protein interactions can be predicted using phylogenetic profiles, which are defined as patterns of gene pairs that are concurrently present or absent across genomes.

**Matrix**  Is a web server that predicts interaction between two protein families (http://orion.icmb.utexas.edu/cgi-bin/matrix/matrix-index.pl).

**Automated Detection and Validation of Interaction Based on the Coevolutions (ADVICE)**  Is a similar web server providing prediction of interacting proteins using the mirror-tree approach (http://advice.i2r.a-star.edu.sg/).

Commercial softwares used for proteomics analysis as follows.

**ChromaTOF**  Is a software used for mass spectrometer data system for acquiring, processing, and reporting data, available at https://www.leco.com/products/separa-tion-science/software-accessories/chromatof-software.

**GeneSpring**  Provides powerful, accessible statistical tools for intuitive data analysis and visualization, designed specifically for the needs of biologists, available at https://www.agilent.com/en/products/software-informatics/life-sciences-informatics/genespring-gx.

**MarkerView**  Is a data visualization program designed for scientists to gain valuable insight into any trends within mass spectral data, available at https://sciex.com/products/software/markerview-software.

**Mass Frontier**  Is a spectral interpretation software, provides small-molecule structural elucidation for research into metabolism, metabolomics, forensics, natural products, impurities, and degradants, available at https://www.thermofisher.com/order/catalog/product/IQLAAEGABOFAGUMZZZ.

## 7.4  Bioinformatics Approaches in Transcriptome

Transcriptome is the study of genome analysis and quantification of RNA which includes mRNAs, noncoding RNAs, and small RNAs in health and disease. There are few public transcriptomic databases, i.e., Gene Expression Omnibus (Barrett et al. 2010), ArrayExpress, and MINISEQE (Brazma 2009). Meta-analysis approach should be applied to decrease data bias and elevate statistical power (Rung and Brazma 2013), which will facilitate transcriptomic information. Many tools are used for transcriptome analysis; a few are mentioned in Table 7.2.

**Table 7.2**  Tools commonly used for transcriptomic analysis

| S. No | Tool name | Description | Link |
|---|---|---|---|
| 1 | SPARTA | Program for automated reference-based bacterial RNA-sequence transcriptome analysis | http://sparta.readthedocs.io/en/latest/ |
| 2 | PIVOT | Interactive analysis and visualization of transcriptomics data | http://kim.bio.upenn.edu/software/pivot.shtml |
| 3 | ReTrOS | Reconstructing transcriptional activity from gene and protein expression data | http://www2.warwick.ac.uk/fac/sci/systemsbiology/research/software/ |
| 4 | WebGIVI | Gene enrichment analysis and visualization tool | http://raven.anr.udel.edu/webgivi/ |
| 5 | SAMSA | Comprehensive metatranscriptome analysis pipeline | http://github.com/transcript/SAMSA |
| 6 | pcaReduce | Hierarchical clustering of single-cell transcriptional profiles | https://github.com/JustinaZ/pcaReduce |
| 7 | GigaTON | New reference transcriptome in the pacific oyster *Crassostrea gigas* | http://www.ncbi.nlm.nih.gov/genome/annotation_euk/Crassostrea_gigas/100/ |

Transcriptome analysis using ESTs, SAGE, and DNA microarrays forms the core of functional genomics and is a key to understanding the interactions of genes and their regulation at the whole genome level.

**Expressed Sequence Tags (ESTs)** One of the high-throughput approaches to genome-wide profiling of gene expression is sequencing, expressed sequence tags (short sequences obtained from cDNA clones and serve as short identifiers of full-length genes). ESTs are typically in the range of 200 to 400 nucleotides in length obtained from either the 5 end or 3 end of cDNA inserts. The EST data are able to provide a rough estimate of genes that are actively expressed in a genome under a particular physiological condition because the frequencies for particular ESTs reflect the abundance of the corresponding mRNA in a cell, which corresponds to the levels of gene expression at that condition. The rapid aggregation of EST sequences has prompted the establishment of public and private databases to archive the data. For example, GenBank has a special EST database, dbEST (www.ncbi. nlm.nih.gov/dbEST/), that contains EST collections for a large number of organisms (>250). EST Index Construction, one of the goals of the EST databases is to organize and consolidate the largely redundant EST data to improve the quality of the sequence information so the data can be used to extract full-length cDNAs. The process includes a preprocessing step that removes vector contaminants and masks repeats. VecScreen can be used to screen out bacterial vector sequences. This is followed by a clustering step that associates EST sequences with unique genes. The next step is to derive consensus sequences by fusing redundant, overlapping ESTs and to correct errors, especially frameshift errors. This step results in longer EST contigs. The procedure is somewhat similar to the genome assembly of shotgun sequence reads. Once the coding sequence is identified, it can be annotated by translating it into protein sequences for database similarity searching. To go another step further, compiled ESTs can be used to align with the genomic sequence if available to identify the genome locus of the expressed gene as well as intron–exon boundaries of the gene. This is usually performed using the program SIM4 (http://pbil.univ-lyon1.fr/sim4.php). The clustering process that reduces the EST redundancy and produces a collection of nonredundant and annotated EST sequences is known as gene index construction. The following lists a couple of major databases that index EST sequences. UniGene (www.ncbi.nlm.nih.gov/UniGene/) is an NCBI EST cluster database. Each cluster is a set of overlapping EST sequences that are computationally processed to represent a single expressed gene. The database is constructed based on combined information from dbEST, GenBank mRNA database, and "electronically spliced" genomic DNA. Only ESTs with 3 poly-A ends are clustered to minimize the problem of chimerism. The resulting 3 EST sequences provide a more unique representation of the transcripts, and errors in individual ESTs are corrected; the sequences are then partitioned into clusters and assembled into contigs. The final result is a set of nonredundant, gene-oriented clusters known as UniGene clusters. Each UniGene cluster represents a unique gene and is further annotated for putative function and its gene locus information, as well as information related to the tissue type where the gene has been expressed. TIGR Gene Indices (www.tigr. org/tdb/tgi.shtml) is an EST database that uses a different clustering method from UniGene. It compiles data from dbEST, GenBank mRNA, and genomic DNA data and TIGR's own sequence database.

**Serial Analysis of Gene Expression (SAGE)** SAGE is a high-throughput, sequence-based approach for global gene expression profile analysis. Unlike EST sampling, SAGE is more quantitative in determining mRNA expression in a cell. In this method, short fragments of DNA (usually 15 base pairs) are excised from cDNA sequences and used as unique markers of the gene transcripts. The sequence fragments are termed tags. They are subsequently concatenated (linked together), cloned, and sequenced. The transcript analysis is carried out computationally in a serial manner. Once gene tags are unambiguously identified, their frequency indicates the level of gene expression. SAGEmap (www.ncbi.nlm.nih.gov/SAGE/) is a SAGE database created by NCBI. Given a cDNA sequence, one can search SAGE libraries for possible SAGE tags and perform "virtual" Northern blots that indicate the relative abundance of a tag in a SAGE library. Each output is hyperlinked to a particular UniGene entry with sequence annotation. SAGE xProfiler (www.ncbi. nlm.nih.gov/SAGE/sagexpsetup.cgi) is a web-based program that allows a "virtual subtraction" of an expression profile of one library (e.g., normal tissue) from another (e.g., diseased tissue). SAGE Genie (http://cgap.nci.nih.gov/SAGE) is another NCBI web-based program that allows matching of experimentally obtained SAGE tags to known genes. It provides an interface for visualizing human gene expression.

**Microarray-Based Approaches** The most commonly used global gene expression profiling method in current genomics research is the DNA microarray-based approach. A microarray (or gene chip) is a slide attached with a high-density array of immobilized DNA oligomers representing the entire genome of the species under study. Each oligomer is spotted on the slide and serves as a probe for binding to a unique, complementary cDNA. The entire cDNA population, labeled with fluorescent dyes or radioisotopes, is allowed to hybridize with the oligoprobes on the chip. The amount of fluorescent or radiolabels at each spot position reflects the amount of corresponding mRNA in the cell. Using this analysis, patterns of global gene expression in a cell can be examined. Sets of genes involved in the same regulatory or metabolic pathways can potentially be identified.

DNA microarrays are generated by fixing oligonucleotides onto a solid support such as a glass slide using a robotic device. The oligonucleotide array slide represents thousands of preselected genes from an organism. The length of oligonucleotides is typically in the range of 25–70 bases long. The oligonucleotides are called probes that hybridize to labeled cDNA samples. Shorter oligoprobes tend to be more specific in hybridization because they are better at discriminating perfect complementary sequences from sequences containing mismatches. However, longer oligos can be more sensitive in binding cDNAs. Many programs have been developed that use these rules in designing probe sequences for microarray spotting. There are various programs for microarray analysis such as the following.

**OligoWiz** OligoWiz is a Java program that runs locally but allows the user to connect to the server to perform analysis via a graphic user interface. It designs oligonucleotides by incorporating multiple criteria including homology, low complexity, and relative position within a transcript. This program is available at (www.cbs.dtu.dk/services/OligoWiz/).

**OligoArray** OligoArray is also a Java client-server program that computes oligonucleotides for microarray construction. It uses the normal criteria with an emphasis on gene specificity and secondary structure for oligonucleotides. This program is available at http://berry.engin.umich.edu/oligoarray2/.

The expression of genes is measured via the signals from cDNAs hybridizing with the specific oligonucleotide probes on the microarray. The cDNAs are obtained by extracting total RNA or mRNA from tissues or cells and incorporating fluorescent dyes in the DNA strands during the cDNA biosynthesis. The colored image is stored as a computer file for further processing.

Image processing is to locate and quantitate hybridization spots and to separate true hybridization signals from background noise and artifacts, as they include nonspecific hybridization, unevenness of the slide surface, and the presence of contaminants such as dust on the surface of the slide. However, there are a small number of free image-processing software programs available on the Internet.

**ArrayDB** Is a web interface program that allows the user to upload data for graphical viewing (http://genome.nhgri.nih.gov/arraydb/). The user can present histograms, select actual microarray slide images, and display detailed information of each spot which is linked to functional annotation of the corresponding gene in the UniGene, Entrez, dbEST, and KEGG databases.

**ScanAlyze** Is a Windows program for microarray fluorescent image analysis. It features semiautomatic spot definition and multichannel pixel and spot analyses (http://rana.lbl.gov/EisenSoftware.html).

**TIGR Spotfinder** Is another Windows program for microarray image processing using the TIFF image format (http://www.tigr.org/softlab/). It uses an adaptive threshold algorithm, which resolves the boundaries of spots according to their shapes.

Following image processing, the digitized gene expression data need to be further processed before differentially expressed genes can be identified. This processing is referred to as data normalization and is designed to correct bias owing to variations in microarray data collection rather than intrinsic biological differences. The following programs are specialized in image analysis and data normalization.

**DNA-Arrays Analysis Tools** It is a web-based program for DNA array data analysis including two sample correlation plots (hierarchical clustering, SOM (self-organizing map), self-organizing hierarchical neural network (SOTA), and various tree viewers (http://bioinfo.cnio.es/).

**TIGR Tm4** A software developed mainly for managing, evaluating, and quantifying for better understanding of data derived from microarray experiment (http://home.cc.umanitoba.ca/~psgendb/birchdoc/package/TIGR-TM4.html).

**Array Designer** Array designer efficiently designs hundreds of specific oligos for single nucleotide polymorphism detection or expression studies or hundreds of polymerase chain reaction primer pairs for cDNA microarrays. The software is available at http://www.premierbiosoft.com/dnamicroarray/.

**ArrayMining** ArrayMining, an online server for automatic microarray analysis, provides information based on feature selection, clustering, and prediction analysis (http://lcsb-repexplore.uni.lu/ASAP/microarrayinfobiotic.php).

**ArrayPlot** ArrayPlot is a Windows program that allows visualization, filtering, and normalization of raw microarray data. It has an interface to view significantly up-regulated or downregulated genes (www.biologie.ens.fr/fr/genetiqu/puces/publications/arrayplot/index.html).

**SNOMAD** Is a web server for microarray data normalization. It provides scatter plots based on raw signal intensities and performs log-transformation and linear regression as well as Lowess regression analysis of the data (http://pevsnerlab.kennedykrieger.org/snomadinput.html).

For statistical analysis to identify differentially expressed genes, the following programs are available.

**MA-ANOVA**  Is a statistical program for Windows and UNIX that uses ANOVA to analyze microarray data. It calculates log ratios, displays ratio-intensity plots, and performs permutation tests and bootstrapping of confidence values (www.jax.org/staff/churchill/labsite/software/anova/).

**Cyber-T**  Is a web server that performs $t$-tests on observed changes of replicate gene expression measurements to identify significantly differentially expressed genes (http://visitor.ics.uci.edu/genex/cybert/).

## 7.5  Bioinformatics Approaches in Metabolomics

Metabolomics aims for analysis of the all metabolites expressed in a biological system (e.g., endogenous or exogenous small molecules) (Psychogios et al. 2011) and is majorly used in various fields (e.g., agriculture, pharma, clinic, environment, and nutrition). Metabolomics has been divided into two distinct approaches, untargeted and targeted metabolomics (Table 7.3).

Metabolomics aims to measure a wide breadth of small molecules in the context of physiological stimuli or in disease states (Roberts et al. 2012). Metabolites help to understand insights of phenotypic expression as they are produced by enzymatic reactions mediating complex biological processes. The two leading analytical approaches to metabolomics are mass spectrometry (MS) and nuclear magnetic resonance (NMR) spectroscopy (Markley et al. 2017). Untargeted metabolomics is a useful approach for the simultaneous analysis of many compounds in herbal products (Commisso et al. 2013). There are numerous tools and databases which are used for metabolomics analysis (Table 7.4).

**Table 7.3**  Major approaches in metabolomics

| S. No | Metabolomics approaches | Description | Application |
|-------|------------------------|-------------|-------------|
| 1 | Untargeted metabolomics | Less specific and sensitive | Diagnostics and drug molecule development |
| | | Intended comprehensive analysis of all the measurable analytes in a sample (chemical unknowns, targeted metabolomics) | Based on information (e.g., stable isotopes and models of metabolic), networks allow identification of the flux through |
| | | | biochemical pathways (Lee et al. 2010) |
| 2 | Targeted metabolomics | Specific and sensitive | Used for identification (based on univariate and multivariate analyses) and then used to search databases (e.g., Kyoto Encyclopedia of Genes and Genomes) (Kanehisa and Goto 2000) |
| | | The measurement of defined groups of chemically identified and biochemically annotated metabolites | |

**Table 7.4** Database/tools used in metabolomics

| S. No | Metabolomics approaches | Database/tools | Description | Links |
|---|---|---|---|---|
| 1 | Metabolomics database | KEGG pathway database | Database provides information regarding pathways, drug molecules, chemical compounds, etc. (Kanehisa and Goto 2000) | https://www.genome.jp/kegg/pathway.html |
| | | MetaCyc | A database used for metabolic analysis (e.g., enzymes, reactions, metabolites, etc.) | https://metacyc.org/ |
| | | EcoCyc | A database that provides information of pathways, metabolites, and genome information of *Escherichia coli* | https://ecocyc.org/ |
| | | BioCyc | A database which associated sequenced genome information with metabolic pathways | https://biocyc.org/ |
| | | Human Metabolome Database (HMDB) | A database that provides data regarding metabolites (small molecule), expressed in the body of *Homo sapiens* | http://www.hmdb.ca/ |
| | | MetabolomeExpress | MetabolomeExpress gives another chance to the general metabolomics network to straightforwardly exhibit online the crude and prepared GC/MS information hidden on their metabolomics publications | https://www.metabolome-express.org/ |
| | | MetaMapR | It coordinates enzymatic changes with structural similarity and mass spectral similarity to create richly associated metabolic networks | http://dgrapov.github.io/MetaMapR/ |
| 2 | NMR based | Normalization and evaluation (NOREVA) | It is an online tool that analyzes various normalization methods derived from MS-based metabolomics information | http://server.idrb.cqu.edu.cn/noreva/ |
| | | NMRProcFlow | A software developed to give a complete set of programs for visualization and processing for data derived from 1-D NMR | https://nmrprocflow.org/ |
| 3 | MS-based untargeted metabolomics | MetaboLights | MetaboLights database is used for metabolomics experiments and derived information and covers metabolite structure and their reference spectra and moreover their biological roles, locations, and concentrations as well | https://www.ebi.ac.uk/metabolights/ |
| | | METLIN | METLIN Metabolomics Database is a repository of metabolite information as well as tandem mass spectrometry data | http://enigma.lbl.gov/metlin/ |
| | | MMCD | Madison Metabolomics Consortium Database (MMCD) is a resource for metabolomics research based on nuclear magnetic resonance (NMR) spectroscopy and mass spectrometry (MS) | www.mmcd.nmrfam.wisc.edu |

# References

Barrett, T., Troup, D. B., Wilhite, S. E., Ledoux, P., Evangelista, C., Kim, I. F., Tomashevsky, M., Marshall, K. A., Phillippy, K. H., Sherman, P. M., & Muertter, R. N. (2010). NCBI GEO: Archive for functional genomics data sets—10 years on. *Nucleic Acids Research, 20:39*(suppl_1), D1005–D1010.

Brazma, A. (2009). Minimum information about a microarray experiment (MIAME)–successes, failures, challenges. *The Scientific World Journal, 9*, 420–423.

Bunnik, E. M., & Le Roch, K. G. (2013). An introduction to functional genomics and systems biology. *Advances in Wound Care, 1;2*(9), 490–498.

Commisso, M., Strazzer, P., Toffali, K., Stocchero, M., & Guzzo, F. (2013). Untargeted metabolomics: An emerging approach to determine the composition of herbal products. *Computational and Structural Biotechnology Journal, 4*(5), e201301007.

Fahrmann, J., Grapov, D., Yang, J., Hammock, B., Fiehn, O., Bell, G. I., & Hara, M. (2015). Systemic alterations in the metabolome of diabetic nod mice delineate increased oxidative stress accompanied by reduced inflammation and hypertriglyceridemia. *American Journal of Physiology. Endocrinology and Metabolism, 308*(11), E978–E989.

Friedrich, N. (2012). Metabolomics in diabetes research. *The Journal of Endocrinology, 215*(1), 29–42.

Gracie, S., Pennell, C., Ekman-Ordeberg, G., et al. (2011). An integrated systems biology approach to the study of preterm birth using -omic technology – A guideline for research. *BMC Pregnancy and Childbirth, 11*, 71.

Harrow, J., Frankish, A., Gonzalez, J. M., et al. (2012). GENCODE: The reference human genome annotation for the ENCODE project. *Genome Research, 22*, 1760–1774.

Kanehisa, M., & Goto. (2000). KEGG: Kyoto encyclopedia of genes and genomes. *Nucleic Acids Research, 28*(1), 27–30.

Kuhn, R. M., Karolchik, D., Zweig, A. S., Wang, T., Smith, K. E., Rosenbloom, K. R., Rhead, B., Raney, B. J., Pohl, A., Pheasant, M., Meyer, L., Hsu, F., Hinrichs, A. S., Harte, R. A., Giardine, B., Fujita, P., Diekhans, M., Dreszer, T., Clawson, H., Barber, G. P., Haussler, D., & Kent, W. J. (2009). The UCSC genome browser database: Update 2009. *Nucleic Acids Research, 37*, D755–D761.

Lee, D. Y., Bowen, B. P., & Northen, T. R. (2010). Mass spectrometry-based metabolomics, analysis of metabolite-protein interactions, and imaging. *BioTechniques, 49*(2), 557–565.

Markley, J. L., Brüschweiler, R., Edison, A. S., Eghbalnia, H. R., Powers, R., Raftery, D., & Wishart, D. S. (2017, February 1). The future of NMR-based metabolomics. *Current Opinion in Biotechnology, 43*, 34–40.

Ono, Y., Asai, K., & Hamada, M. (2013). PBSIM: PacBio reads simulator--toward accurate genome assembly. *Bioinformatics, 29*, 119–121.

Psychogios, N., Hau, D. D., Peng, J., Guo, A. C., Mandal, R., Bouatra, S., Sinelnikov, I., Krishnamurthy, R., Eisner, R., Gautam, B., Young, N., Xia, J., Knox, C., Dong, E., Huang, P., Hollander, Z., Pedersen, T. L., Smith, S. R., Bamforth, F., Greiner, R., McManus, B., Newman, J. W., Goodfriend, T., & Wishart, D. S. (2011). The human serum metabolome. *PLoS One, 6*(2), e16957.

Roberts, L. D., Souza, A. L., Gerszten, R. E., & Clish, C. B. (2012). Targeted metabolomics. *Current Protocols in Molecular Biology, CHAPTER*, Unit 30.2. https://doi.org/10.1002/0471142727. mb3002s98.

Rung, J., & Brazma, A. (2013). Reuse of public genome-wide gene expression data. *Nature Reviews. Genetics, 14*(2), 89–99.

Simpson, J. T., & Pop, M. (2015). The theory and practice of genome sequence assembly. *Annual Review of Genomics and Human Genetics, 16*, 153–172.

Venter, J. C., Smith, H. O., & Adams, M. D. (2015). The sequence of the human genome. *Clinical Chemistry, 61*, 1207–1208.

Wikoff, W. R., Grapov, D., Fahrmann, J. F., DeFelice, B., Rom, W., Pass, H., Kim, K., Nguyen, U., Taylor, S. L., Kelly, K., & Fiehn, O. (2015). Metabolomic markers of altered nucleotide metabolism in early stage adenocarcinoma. *Cancer Prevention Research (Philadelphia, Pa.), 8*(5), 410–418.

Zhou, X., Peris, D., Kominek, J., Kurtzman, C. P., Hittinger, C. T., & Rokas, A. (2016, November 1). In silico whole genome sequencer and analyzer (iWGS): A computational pipeline to guide the design and analysis of de novo genome sequencing studies. *G3: Genes, Genomes, Genetics, 6*(11), 3655–62.